Textbook of
EARTH SCIENCE

Textbook of EARTH SCIENCE

Dr. C.D. Kathuria

CENTRUM PRESS
NEW DELHI-110002 (INDIA)

CENTRUM PRESS
H.O.: 4360/4, Ansari Road, Daryaganj,
New Delhi-110002 (India)
Tel: 23278000, 23261597, 23255577, 23286875
B.O.: No. 1015, Ist Main Road, BSK IIIrd Stage,
IIIrd Phase, IIIrd Block, Bengaluru-560085 (INDIA)
Tel: 080-41723429
Email: centrumpress@gmail.com
Visit us at: www.centrumpress.com

Textbook of Earth Science

First Edition, 2011

ISBN 978-93-81293-46-1

PRINTED IN INDIA

Printed at Suman Printers, Jagatpuri Extn., Delhi

Contents

Preface

Earth Science forces, including earthquakes, fossils, the Earth's spheres, the Big Bang Theory, El Nino, and the atmosphere, are presented at a level appropriate for beginners. Readers will also learn about more advanced topics within earth science, including geochronology, seismology, and geomagnetism.

Earth science, is an all-embracing term for the sciences related to the planet Earth. It is arguably a special case in planetary science, the Earth being the only known life-bearing planet. There are both reductionist and holistic approaches to Earth sciences. The formal discipline of Earth sciences may include the study of the atmosphere, oceans and biosphere, as well as the solid earth. Typically Earth scientists will use tools from physics, chemistry, biology, chronology and mathematics to build a quantitative understanding of how the Earth system works, and how it evolved to its current state.

The effects of organisms can and has affected Earth's atmosphere. In the earliest days of Earth's history, the atmosphere did not have much free oxygen. The Earth's oxygen was emitted as waste by colonial bacteria while they built huge colonial masses now called "stromatolites." Alternating layers of fine grained silica with iron oxides in several phases suggests winter clay or fine quartz runoff being recolonized every year by new iron-fixing bacteria. Their huge and deep deposits are mined for much industrial iron. Areas such as Iron Mountain, Michigan and areas in Western Australia have huge underground and open pit operations. The organisms, however, changed the atmosphere and most were driven far underwater and underground where their descendents are in some places, still the dominant force on their changed Earth.

Earth is blanketed by an atmosphere consisting of 78.0% nitrogen, 20.9% oxygen, and 1% Argon. The atmosphere has five layers: troposphere, stratosphere, mesosphere, thermosphere, and exosphere; and 75% of the atmosphere's gases are in the bottom-most layer, the troposphere. The magnetic field created by mantle's internal motions produces the magnetosphere which protects the Earth's atmosphere from the solar wind. It is theorized that the solar wind would strip away earth's atmosphere in a few million years were it not for the Earth's electromagnet. And since earth is 4.5 billion years old, earth would not have an atmosphere by now if there were no magnetosphere.

Earth science generally recognizes four spheres, the lithosphere, the hydrosphere, the atmosphere, and the biosphere; these correspond to rocks, water, air, and life. Some practitioners include, as part of the spheres of the Earth, the cryosphere (corresponding to ice) as a distinct portion of the hydrosphere, as well as the pedosphere (corresponding to soil) as an active and intermixed sphere.

The Earth Science India (ESI) was launched during the international year of planet earth with a view to enhance communication between earth scientists of India and their peers from other parts of the world. This book is intended for the student. It provides all the basic concepts and latest information on geology. As the phenomena of evolution is on run so all the relevant fields are affected and ever changing.

—Editor

1

Introduction

The Universe has existed for about 15 billion years.

- In it, the speed of light is always faster than speed of sound.
- Therefore the first event generated a large flash of light outwards in all directions long before it would have made a big, or any sort of bang.
- Stars are suns. Billions and billions of suns (thank you, Carl Sagan).
- Galaxies are collections of suns. Our sun is in a galaxy which we see edge on as the Milky Way.
- The sun's average temperature is about 5480 degrees Celsius, or 548 times boiling water.

Our sun has a Solar System which includes Mercury, Venus, Earth, Mars, the asteroid belt, Jupiter, Saturn, Neptune, Uranus, Pluto/Charon and the Oort Cloud of comets. There is a regular spacing to the planets. Using the Sun-Mercury distance as one unit, the spacing is 1:1:2:3:5:8 and so on, more or less and excluding Pluto/Charon. The age of the Earth and Moon are about the same; about 3.9 billion years.

Earth is about 25,000 miles around. This was first calculated by Eratosthenes, a teacher at the Alexandria Museum about 239 BC. He assumed that the sun's rays were parallel and that the sun was a great distance away. An ancient well is said to have reflected the light directly upward to light a monument and was quite a tourist attraction for several thousand years. So everyone knew that the sun stood overhead in Aswan on

Summer Solstice. Eratosthenes measured the angle of the sun at noon in Alexandria and found it was about 7.5 degrees out of a total of 360.

Since 7.5 divided by 360 degrees in a circle works out to about 1/50 of a circle, he now knew the angle of the Sun's rays between Alexandria and Aswan. The Egyptian post office is said to have provided the ground distance between the two points as the equivalent of 500 miles. So if 500 miles is 1/50th of a circle, the whole circle is 25,000 miles around. Almost spot on, as it turns out. From this, he calculated that it is about 8,000 miles to the centre of the Earth.

But his accomplishment is dwarfed by Mesopotamian astronomers who described the following systems almost exactly, but in ancient Sumerican terms as early as 2,500 BC and copied and recopied as "religious texts" for centuries afterward along with the construction mathematics to build observatories and instruments.

It has been known since ancient times that:

- The North Pole appears to point to a north star. Right now it's a star called Polaris. Stars rotate around Polaris; some are always visible at night, others appear to rise and set. Stars which rotate closest to Polaris cannot be seen from the Southern Hemisphere.
- In the Southern Hemisphere, the Southern Cross points toward the pole, but there is no star exactly there right now. Starts rotate around the southern pole; some are always visible at night, others appear to rise and set. Starts which rotate closest to the southern pole cannot be seen from the Northern Hemisphere.
- The stars which rise and set can be seen from both hemispheres.
- The plane in which most of the planets in our Solar System rotates is called the plane of the ecliptic. The zodiac constellations mark the plane of the ecliptic and the planets, the moon and the sun all appear to rise and set in this plane.
- The moon revolves around the Earth once ~ every 28 days and rotates on its axis once in the same time; keeping its face always towards the Earth.

Day	*Phase*	*Colour*	*Rising time*
1	Full	white	at or just after sunset
7	Last Quarter	half black/half white	near midnight
14	New Moon	black	near sunrise
21	First Quarter	half white/half black	near noon

- The Earth's axis tilts about 23.5 degrees now. It wobbles over about 26,000 years in a full circle so that while Polaris is the North Star at the present time, the north polar star changes slowly over time and will return to being Polaris again in about another 26,000 years.
- We know that our axis points at Polaris-on a human life span and that the Sun is in the middle of the Solar System.
- We know that there is a solid core in the middle of the Earth which rotates within a liquid outer core very rapidly and generates the Earth's magnetic field. The axis of the solid core is not the same as the axis of the Earth; it moves, but returns to about the same point on Earth's surface once every 400 years. This is why the magnetic pole and the axial pole are not the same.
- The Earth's orbit is an ellipse, not a circle. It is closest to the sun around January 3 (perihelion) and farthest from the sun around July 3 (aphelion). The difference is minimal. The distance is about 93 million miles.
- Our seasons are a function of the axial tilt. In Northern Hemisphere summer, the north pole receives 24 hours a day of sunlight for 6 weeks, warming the whole hemisphere. At that time, the Southern Hemisphere is having winter and the south pole is having 24 hours a day of darkness for the same 6 weeks. In Spring and Fall-around the equinoxes, the Earth is evenly heated and every where on Earth has 12 hours of day and 12 hours of night that day. Hence the name, Equinox. The winter effects on each Hemisphere are the opposite of the summer effects.
- The tilt of the Earth is also responsible for the length of the day changing throughout the year. In the northern hemisphere, the longest day in the is Summer Solstice,

about June 21 and the shortest day is Winter Solstice about December 21 with the Equinoxes at March and September 21 or so every year.

- The tilt is also responsible for the apparent change in where the sun appears to rise. In Northern Hemisphere summer, the sun rises far to the northeast and sets far to the northwest; in winter it appears to rise southeast and set southwest. At the equinoxes, it appears to rise due east and set due west.
- The moon orbits the Earth between five degrees north and south of the equator, crossing that line twice a year. That is why in Northern Hemisphere summer, the moon appears to ride low in the sky (it is in the southern hemisphere itself) and in Northern winter, the moon appears high in the sky. It is the opposite of the sun which rides high in summer and low in winter.

Here are some other factoids about our planet:

- Earth rotates on its axis about once every 24 hours and revolves around the sun about once every 365 days. The speed at the equator is about 1040 miles per hour.
- Northern Hemisphere is 50 percent of the land — the Southern is 75 percent ocean.
- Average land height is 0.8 kilometres
- Average ocean depth 3.8 kilometres (3800 meters)
- Lowest point-Marianas Trench 11,000 meters
- Highest point-Mt. Everest 8936 meters
- Molten magma, hot rock in Hawaii'an volcanos can reach 3000 C or 300 times boiling water.
- Before artificial lighting, it was possible to see sun shadow, moon shadow and the shadow of Venus and bows on both the sun (rainbows) and moon under certain atmospheric conditions.

The direction of Earth's spin controls the directions of winds and waves.

- At the Equator to 30 degrees N or S, the winds run away from the spin (easterlies).
- In the mid-latitudes from 30 to 60 degrees N or S, the winds run toward the spin (westerlies).

- In the polar latitudes, the winds again run away from the spin (easterlies).
- The direction of the prevailing winds influences currents which flow away from the spin at the Equator and are deflected by continents towards the respective poles and back around to their place of origin.

Water is amazing stuff.

- Unlike just about every other solid, the solid form of water (ice) floats which means it has a density less than that of its liquid phase.
- Fresh water freezes at O Celsius (32 F) and boils at 100 C (212 F) at sea level. In both cases, its density becomes less than one.
- Sea water is more dense than freshwater. The average density of seawater 1.026 g/cm^3 while the average density of water 1.0 g/cm^3.
- Most water on Earth is salty (97 percent) leaving just 3 percent in glaciers and the other 1 percent in streams, lakes, groundwater and water vapour.

Pressure can do some unbelievable things.

- Changing the pressure of a substance can change its temperature. Compression raises temperature, expansion lowers temperature.
- The pressure of air at sea level is called 1 atm (about 100 g/cc or 14.7 lb/sq.in or 29.92 inches of mercury). The instrument to measure barometric pressure gave rise to the terms "low pressure" and "high pressure." The air at sea level, being the most compressed, is usually the warmest and air gets colder as you go up. Half the air is in the bottom five miles of the 150 mile thick atmosphere.
  * At the surface of the earth, water boils at 100 degrees Celsius (212F) and freezes at 0 C (32F).
  * At higher elevations, water actually boils at a lower temperature, but takes longer to do it due to lower air pressures.
- Confining pressure in the ocean rises 1 atm per 10 meters of depth + the 1 atm of air at the surface, so the

pressure of air at the average ocean depth of 3800 meters is 381 atmospheres. Temperature at the bottom is variable and can be quite high near underseas volcanos.

- Confining pressure in rock results in rock pressure increases of about 250-300 bars/kilometre and increases of around 30 degrees C/kilometer. Temperature, however, can vary based on nearness to volcanic activity.
- Fluids tend to follow faults or cracks in rocks, upward towards areas of lesser pressure.
- Minerals form in places with abrupt changes in pressure/temperature such as the air/water interface.

The U.S. should have already gone on the metric system. It would make it easier to teach because we wouldn't have to convert all the units between the two systems all the time. Learn to use metric.

- 1 kilometre equals 0.62 of a mile; one mile equals about one and a half kilometres (1.6 km)
- 1 inch is equal to 2.54 centimeters, therefore each centimetre is about one half inch.
- 1 pint is equal to about one pound, one gallon of water weighs about 8 pounds
- one kilogram of water equals one liter of water, also 2.24 pounds
- there are, therefore 2 and one fourth pints in a liter
- One gram of fresh water is equal to one cubic centimetre of water and also equal to one millilitre of water (1/1000 liter) at 20 degrees C (room temperature). Fresh water has a density of one gram/cc or 1.
- Lead is one of the most dense materials on Earth. It has a density of about 12 g/cc.

The effects of organisms can and has affected Earth's atmosphere. In the earliest days of Earth's history, the atmosphere did not have much free oxygen. The Earth's oxygen was emitted as waste by colonial bacteria while they built huge colonial masses now called "stromatolites." Alternating layers of fine grained silica with iron oxides in several phases suggests winter clay or fine quartz runoff being recolonized every year by new iron-fixing bacteria. Their huge and deep deposits are mined for much industrial iron. Areas such as Iron Mountain, Michigan and areas in Western Australia have huge underground and open pit operations. The organisms, however, changed the atmosphere and most were driven far underwater and underground where their descendents are in some places, still the dominant force on their changed Earth.

EARTH SCIENCE

Earth science (also known as geoscience, the geosciences or the Earth sciences), is an all-embracing term for the sciences related to the planet Earth. It is arguably a special case in planetary science, the Earth being the only known life-bearing planet.

There are both reductionist and holistic approaches to Earth sciences. The formal discipline of Earth sciences may include the study of the atmosphere, oceans and biosphere, as well as the solid earth. Typically Earth scientists will use tools from physics, chemistry, biology, chronology and mathematics to build a quantitative understanding of how the Earth system works, and how it evolved to its current state.

Fields of Study

The following fields of science are generally categorized within the geosciences:

- Geology describes the rocky parts of the Earth's crust (or lithosphere) and its historic development. Major subdisciplines are mineralogy and petrology, geochemistry, geomorphology, paleontology, stratigraphy, structural geology, engineering geology and sedimentology.
- Geophysics and Geodesy investigate the shape of the Earth, its reaction to forces and its magnetic and gravity fields. Geophysicists explore the Earth's core and mantle as well as the tectonic and seismic activity of the lithosphere.
- Soil science covers the outermost layer of the Earth's crust that is subject to soil formation processes (or pedosphere). Major subdisciplines include edaphology and pedology.

- Oceanography and hydrology (includes limnology) describe the marine and freshwater domains of the watery parts of the Earth (or hydrosphere). Major subdisciplines include hydrogeology and physical, chemical, and biological oceanography.
- Glaciology covers the icy parts of the Earth (or cryosphere).
- Atmospheric sciences cover the gaseous parts of the Earth (or atmosphere) between the surface and the exosphere (about 1000 km). Major subdisciplines are meteorology, climatology, atmospheric chemistry and atmospheric physics.
- A very important linking sphere is the biosphere, the study of which is biology. The biosphere consists of all forms of life, from single-celled organisms to pine trees to people. The interactions of Earth's other spheres-lithosphere/geosphere, hydrosphere, atmosphere and/or cryosphere and pedosphere-create the conditions that can support life.

EARTH'S INTERIOR

Plate tectonics, mountain ranges, volcanoes, and earthquakes are geological phenomena that can be explained in terms of energy transformations in the Earth's crust. Beneath the Earth's crust lies the mantle which is heated by the radioactive decay of heavy elements. The mantle is not quite solid and consists of magma which is in a state of semi-perpetual convection. This convection process causes the lithospheric plates to move, albeit slowly. The resulting process is known as plate tectonics.

Plate tectonics might be thought of as the process by which the earth is resurfaced. Through a process called *spreading ridges* (or seafloor spreading), new earth crust is created by the flow of magma from underneath the lithosphere to the surface, through fissures, where it cools and solidifies. Through a process called subduction, crust is pushed underground—beneath the rest of the lithosphere—where it comes into contact with magma and melts—rejoining the mantle from which it originally came. Areas of the crust where new crust is created are called *divergent boundaries*, and areas of the crust where it is brought back into the earth are called *convergent boundaries*. Earthquakes result from the movement of the lithospheric plates, and they often occur near covergent boundaries where parts of the crust are forced into the earth as part of subduction.

Volcanoes result primarily from the melting of subducted crust material. Crust material that is forced into the Asthenosphere melts, and some portion of the melted material becomes light enough to rise to the surface—giving birth to volcanoes.

EARTH'S ELECTROMAGNETIC FIELD

An electromagnet is a magnet that is created by a current that flows around a soft-iron core. Earth has a soft iron core surrounded by semi-liquid materials from the mantle that move in continuous currents around the core; therefore, the earth is an electromagnet. This is referred to as the dynamo theory of Earth's magnetism. The fact that Earth is an electromagnet helps it maintain an atmosphere suitable for life.

ATMOSPHERE

Earth is blanketed by an atmosphere consisting of 78.0% nitrogen, 20.9% oxygen, and 1% Argon. The atmosphere has five layers: troposphere, stratosphere, mesosphere, thermosphere, and exosphere; and 75% of the atmosphere's gases are in the bottom-most layer, the troposphere. The magnetic field created by mantle's internal motions produces the magnetosphere which protects the Earth's atmosphere from the solar wind. It is theorized that the solar wind would strip away earth's atmosphere in a few million years were it not for the Earth's electromagnet. And since earth is 4.5 billion years old, earth would not have an atmosphere by now if there were no magnetosphere.

The atmosphere is composed of 78% nitrogen and 21% oxygen. The remaining one percent contains small amounts of other gases including CO_2 and water vapors. Water vapors and CO_2 allow the Earth's atmosphere to catch and hold the sun's energy through a phenomenon called the greenhouse effect. This allows earth's surface to be warm enough to have liquid water and support life. In addition to storing heat, the atmosphere also protects living organisms by shielding the

Earth's surface from cosmic rays. Note that the level of protection is high enough to prevent cosmic rays from destroying all life on Earth, yet low enough to aid the mutations that have an important role in pushing forward diversity in the biosphere.

Methodology

Like all other scientists, Earth scientists apply the scientific method. They formulate hypotheses after observing events and gathering data about natural phenomena, and then they test hypotheses from such data.

A contemporary idea within earth science is uniformitarianism. Uniformitarianism says that "ancient geologic features are interpreted by understanding active processes that are readily observed". Simply stated, this means that features of the Earth can be explained by the actions of gradual processes operating over long periods of time; for example, a mountain need not be thought of as having been created in a moment, but instead it may be seen as the result of continuous subduction, causing magma to rise and form continental volcanic arcs.

EARTH'S SPHERES

Earth science generally recognizes four spheres, the lithosphere, the hydrosphere, the atmosphere, and the biosphere; these correspond to rocks, water, air, and life. Some practitioners include, as part of the spheres of the Earth, the cryosphere (corresponding to ice) as a distinct portion of the hydrosphere, as well as the pedosphere (corresponding to soil) as an active and intermixed sphere.

Partial List of the Major Earth Science Topics

Atmosphere

- Atmospheric chemistry
- Climatology
- Meteorology
    * Hydrometeorology
- Paleoclimatology.

Biosphere

- Biogeography

- Paleontology
  - * Palynology
  - * Micropaleontology
- Geomicrobiology
- Geoarchaeology.

Hydrosphere

- Hydrology
  - * Limnology
- Hydrogeology
- Oceanography
  - * Chemical oceanography
  - * Marine biology
- Marine geology
  - * Paleoceanography
  - * Physical oceanography.

Lithosphere or Geosphere

- Geology
  - * Economic geology
  - * Engineering geology
  - * Environmental geology
  - * Historical geology
    - — Quaternary geology
  - * Planetary geology
  - * Sedimentology
  - * Stratigraphy
  - * Structural geology
- Geography
  - * Physical geography
- Geochemistry
- Geomorphology
- Geophysics
  - * Geochronology
  - * Geodynamics
  - * Geomagnetism

    * Gravimetry (also part of Geodesy)
    * Seismology
- Glaciology
- Hydrogeology
- Mineralogy
    * Crystallography
    * Gemology
- Petrology
- Speleology
- Volcanology.

Pedosphere

- Soil science
    * Edaphology
    * Pedology.

Systems

- Environmental science
- Geography
    * Human geography
    * Physical geography
- Gaia hypothesis.

Others

- Cartography
- Geoinformatics (GIS)
- Geostatistics
- Geodesy and Surveying
- NASA Earth Science Enterprise.

2

The Earth

Earth (or the Earth) is the third planet from the Sun, and the densest and fifth-largest of the eight planets in the Solar System. It is also the largest of the Solar System's four terrestrial planets. It is sometimes referred to as the World, the Blue Planet, or by its Latin name, *Terra*.

Home to millions of species including humans, Earth is currently the only place where life is known to exist. The planet formed 4.54 billion years ago, and life appeared on its surface within a billion years. Earth's biosphere has significantly altered the atmosphere and other abiotic conditions on the planet, enabling the proliferation of aerobic organisms as well as the formation of the ozone layer which, together with Earth's magnetic field, blocks harmful solar radiation, permitting life on land. The physical properties of the Earth, as well as its geological history and orbit, have allowed life to persist during this period. The planet is expected to continue supporting life for at least another 500 million years.

Earth's outer surface is divided into several rigid segments, or tectonic plates, that migrate across the surface over periods of many millions of years. About 71% of the surface is covered with salt water oceans, the remainder consisting of continents and islands which together have many lakes and other sources of water contributing to the hydrosphere. Liquid water, necessary for all known life, is not known to exist in equilibrium on any other planet's surface. Earth's poles are mostly covered with solid ice (Antarctic ice sheet) or sea ice (Arctic ice cap). The planet's interior remains active, with a thick layer of relatively solid mantle, a liquid outer core that generates a magnetic field, and a solid iron inner core.

Earth interacts with other objects in space, especially the Sun and the Moon. At present, Earth orbits the Sun once for every roughly 366.26 times it rotates about its axis, which is equal to 365.26 solar days, or one sidereal year. The Earth's axis of rotation is tilted 23.4° away from the perpendicular to its orbital plane, producing seasonal variations on the planet's surface with a period of one tropical year (365.24 solar days). Earth's only known natural satellite, the Moon, which began orbiting it about 4.53 billion years ago, provides ocean tides, stabilizes the axial tilt and gradually slows the planet's rotation. Between approximately 3.8 billion and 4.1 billion years ago, numerous asteroid impacts during the Late Heavy Bombardment caused significant changes to the greater surface environment. Both the mineral resources of the planet, as well as the products of the biosphere, contribute resources that are used to support a global human population. These inhabitants are grouped into about 200 independent sovereign states, which interact through diplomacy, travel, trade, and military action. Human cultures have developed many views of the planet, including personification as a deity, a belief in a flat Earth or in Earth as the centre of the universe, and a modern perspective of the world as an integrated environment that requires stewardship.

CHRONOLOGY

Scientists have been able to reconstruct detailed information about the planet's past. The earliest dated Solar System material was formed 4.5672 ± 0.0006 billion years ago, and by 4.54 billion years ago (within an uncertainty of 1%) the Earth and the other planets in the Solar System had formed out of the solar nebula—a disk-shaped mass of dust and gas left over from the formation of the Sun. This assembly of the Earth through accretion was thus largely completed within 10–20 million years. Initially molten, the outer layer of the planet Earth cooled to form a solid crust when water began accumulating in the atmosphere. The Moon formed shortly thereafter, 4.53 billion years ago. The current consensus model for the formation of the Moon is the giant impact hypothesis, in which the Moon was created when a Mars-sized object (sometimes called Theia) with about 10% of the Earth's mass impacting the Earth in a glancing blow. In this model, some of this object's mass would have merged with the Earth and a portion would have been ejected into space, but enough material would have been sent into orbit to coalesce into the Moon.

Outgassing and volcanic activity produced the primordial atmosphere of the Earth. Condensing water vapour, augmented by ice and liquid water delivered by asteroids and the larger proto-planets, comets, and trans-Neptunian objects produced the oceans. The newly formed Sun was only 70% of its present luminosity, yet evidence shows that the early oceans remained liquid—a contradiction dubbed the faint young Sun paradox. A combination of greenhouse gases and higher levels of solar activity served to raise the Earth's surface temperature, preventing the oceans from freezing over. By 3.5 billion years ago, the Earth's magnetic field was established, which helped prevent the atmosphere from being stripped away by the solar wind. Two major models have been proposed for the rate of continental growth: steady growth to the present-day and rapid growth early in Earth history. Current research shows that the second option is most likely, with rapid initial growth of continental crust followed by a long-term steady continental area. On time scales lasting hundreds of millions of years, the surface continually reshaped as continents formed and broke up. The continents migrated across the surface, occasionally combining to form a supercontinent. Roughly 750 million years ago (Ma), one of the earliest known supercontinents, Rodinia, began to break apart. The continents later recombined to form Pannotia, 600–540 Ma, then finally Pangaea, which broke apart 180 Ma.

EVOLUTION OF LIFE

At present, Earth provides the only example of an environment that has given rise to the evolution of life. Highly energetic chemistry is believed to have produced a self-replicating molecule around 4 billion years ago and half a billion years later the last common ancestor of all life existed. The development of photosynthesis allowed the Sun's energy to be harvested directly by life forms; the resultant oxygen accumulated in the atmosphere and formed a layer of ozone (a form of molecular oxygen [O_3]) in the upper atmosphere. The incorporation of smaller cells within larger ones resulted in the development of complex cells called eukaryotes. True multicellular organisms formed as cells within colonies became increasingly specialized. Aided by the absorption of harmful ultraviolet radiation by the ozone layer, life colonized the surface of Earth. Since the 1960s, it has been hypothesized that severe glacial action between 750 and 580 Ma, during the Neoproterozoic, covered much of the planet in a sheet of ice. This hypothesis has been termed "Snowball Earth", and is of particular interest because it preceded the Cambrian explosion, when multicellular life forms began to proliferate.

Following the Cambrian explosion, about 535 Ma, there have been five major mass extinctions. The most recent such event was 65 Ma, when an asteroid impact triggered the extinction of the (non-avian) dinosaurs and other large reptiles, but spared some small animals such as mammals, which then resembled shrews. Over the past 65 million years, mammalian life has diversified, and several million years ago an African ape-like animal such as *Orrorin tugenensis* gained the ability to stand upright. This enabled tool use and encouraged communication that provided the nutrition and stimulation needed for a larger brain, which allowed the evolution of the human race. The development of agriculture, and then civilization, allowed humans to influence the Earth in a short time span as no other life form had, affecting both the nature and quantity of other life forms. The present pattern of ice ages began about 40 Ma and then intensified during the Pleistocene about 3 Ma. High-latitude regions have since undergone repeated cycles of glaciation and thaw, repeating every 40–100,000 years. The last continental glaciation ended 10,000 years ago.

Future

The future of the planet is closely tied to that of the Sun. As a result of the steady accumulation of helium at the Sun's core, the star's total luminosity will slowly increase. The luminosity of the Sun will grow by 10% over the next 1.1 Gyr

(1.1 billion years) and by 40% over the next 3.5 Gyr. Climate models indicate that the rise in radiation reaching the Earth is likely to have dire consequences, including the loss of the planet's oceans.

The Earth's increasing surface temperature will accelerate the inorganic CO_2 cycle, reducing its concentration to levels lethally low for plants (10 ppm for C4 photosynthesis) in approximately 500 million to 900 million years. The lack of vegetation will result in the loss of oxygen in the atmosphere, so animal life will become extinct within several million more years. After another billion years all surface water will have disappeared and the mean global temperature will reach 70 °C (158 °F). The Earth is expected to be effectively habitable for about another 500 million years from that point, although this may be extended up to 2.3 billion years if the nitrogen is removed from the atmosphere. Even if the Sun were eternal and stable, the continued internal cooling of the Earth would result in a loss of much of its CO_2 due to reduced volcanism, and 35% of the water in the oceans would descend to the mantle due to reduced steam venting from mid-ocean ridges.

The Sun, as part of its evolution, will become a red giant in about 5 Gyr. Models predict that the Sun will expand out to about 250 times its present radius, roughly 1 AU (150,000,000 km). Earth's fate is less clear. As a red giant, the Sun will lose roughly 30% of its mass, so, without tidal effects, the Earth will move to an orbit 1.7 AU (250,000,000 km) from the Sun when the star reaches it maximum radius. The planet was therefore initially expected to escape envelopment by the expanded Sun's sparse outer atmosphere, though most, if not all, remaining life would have been destroyed by the Sun's increased luminosity (peaking at about 5000 times its present level). However, a 2008 simulation indicates that Earth's orbit will decay due to tidal effects and drag, causing it to enter the red giant Sun's atmosphere and be vaporized. Composition and Structure

Earth is a terrestrial planet, meaning that it is a rocky body, rather than a gas giant like Jupiter. It is the largest of the four solar terrestrial planets in size and mass. Of these four planets, Earth also has the highest density, the highest surface gravity, the strongest magnetic field, and fastest rotation. It also is the only terrestrial planet with active plate tectonics.

Shape

The shape of the Earth is very close to that of an oblate spheroid, a sphere flattened along the axis from pole to pole such that there is a bulge around the equator. This bulge results from the rotation of the Earth, and causes the diameter at the equator to be 43 km larger than the pole to pole diameter. The average diameter of the reference spheroid is about 12,742 km, which is approximately 40,000 km/π, as the meter was originally defined as 1/10,000,000 of the distance from the equator to the North Pole through Paris, France.

Local topography deviates from this idealized spheroid, though on a global scale, these deviations are very small: Earth has a tolerance of about one part in about 584, or 0.17%, from the reference spheroid, which is less than the 0.22% tolerance allowed in billiard balls. The largest local deviations in the rocky surface of the Earth are Mount Everest (8848 m above local sea level) and the Mariana Trench (10,911 m below local sea level). Because of the equatorial bulge, the surface locations farthest from the centre of the Earth are the summits of Mount Chimborazo in Ecuador and Huascarán in Peru.

Chemical composition of the crust

Compound	*Formula*	*Composition*	
		Continental	*Oceanic*
silica	SiO_2	60.2%	48.6%
alumina	Al_2O_3	15.2%	16.5%
lime	Ca*	5.5%	12.3%
magnesia	Mg*	3.1%	6.8%
iron(II) oxide	Fe*	3.8%	6.2%
sodium oxide	Na_2*	3.0%	2.6%
potassium oxide	K_2*	2.8%	0.4%
iron(III) oxide	Fe_2O_3	2.5%	2.3%
water	H_2*	1.4%	1.1%
carbon dioxide	CO_2	1.2%	1.4%
titanium dioxide	TiO_2	0.7%	1.4%
phosphorus pentoxide	P_2O_5	0.2%	0.3%
Total		99.6%	99.9%

Chemical Composition

The mass of the Earth is approximately 5.98×10^{24} kg. It is composed mostly of iron (32.1%), oxygen (30.1%), silicon (15.1%), magnesium (13.9%), sulfur (2.9%), nickel (1.8%), calcium (1.5%), and aluminium (1.4%); with the remaining 1.2% consisting of trace amounts of other elements. Due to mass segregation, the core region is believed to be primarily composed of iron (88.8%), with smaller amounts of nickel (5.8%), sulfur (4.5%), and less than 1% trace elements.

The geochemist F. W. Clarke calculated that a little more than 47% of the Earth's crust consists of oxygen. The more common rock constituents of the Earth's crust are nearly all oxides; chlorine, sulfur and fluorine are the only important exceptions to this and their total amount in any rock is usually much less than 1%. The principal oxides are silica, alumina, iron oxides, lime, magnesia, potash and soda. The silica functions principally as an acid, forming silicates, and all the commonest minerals of igneous rocks are of this nature. From a computation based on 1,672 analyses of all kinds of rocks, Clarke deduced that 99.22% were composed of 11 oxides. All the other constituents occur only in very small quantities.

Internal Structure

The interior of the Earth, like that of the other terrestrial planets, is divided into layers by their chemical or physical (rheological) properties, but unlike the other terrestrial planets, it has a distinct outer and inner core. The outer layer of the Earth is a chemically distinct silicate solid crust, which is underlain by a highly viscous solid mantle. The crust is separated from the mantle by the Mohorovièiæ discontinuity, and the thickness of the crust varies: averaging 6 km under the oceans and 30–50 km on the continents. The crust and the cold, rigid, top of the upper mantle are collectively known as the lithosphere, and it is of the lithosphere that the tectonic plates are comprised. Beneath the lithosphere is the asthenosphere, a relatively low-viscosity layer on which the lithosphere rides. Important changes in crystal structure within the mantle occur at 410 and 660 kilometres below the surface, spanning a transition zone that separates the upper and lower mantle. Beneath the mantle, an extremely low viscosity liquid outer core lies above a solid inner core. The inner core may rotate at a slightly higher angular velocity than the remainder of the planet, advancing by 0.1–0.5° per year.

Geologic Layers of the Earth

Depth km	*Component Layer*	*Density g/cm³*
0–60	Lithosphere	—
0–35	Crust	2.2–2.9
35–60	Upper mantle	3.4–4.4
35–2890	Mantle	3.4–5.6
100–700	Asthenosphere	—
2890–5100	Outer core	9.9–12.2
5100–6378	Inner core	12.8–13.1

Heat

Earth's internal heat comes from a combination of residual heat from planetary accretion (about 20%) and heat produced through radioactive decay (80%). The major heat-producing isotopes in the Earth are potassium-40, uranium-238, uranium-235, and thorium-232. At the centre of the planet, the temperature may be up to 7,000 K and the pressure could reach 360 GPa. Because much of the heat is provided by radioactive decay, scientists believe that early in Earth history, before isotopes with short half-lives had been depleted, Earth's heat production would have been much higher. This extra heat production, twice present-day at approximately 3 billion years ago, would have increased temperature gradients within the Earth, increasing the rates of mantle convection and plate tectonics, and allowing the production of igneous rocks such as komatiites that are not formed today.

Present-day major heat-producing isotopes

Isotope	Heat release W/kg isotope	Half-life years	Mean mantle concentration kg isotope/kg mantle	Heat release W/kg mantle
^{238}U	9.46×10^{-5}	4.47×10^{9}	30.8×10^{-9}	2.91×10^{-12}
^{235}U	5.69×10^{-4}	7.04×10^{8}	0.22×10^{-9}	1.25×10^{-13}
^{232}Th	2.64×10^{-5}	1.40×10^{10}	124×10^{-9}	3.27×10^{-12}
^{40}K	2.92×10^{-5}	1.25×10^{9}	36.9×10^{-9}	1.08×10^{-12}

Total heat loss from the Earth is 4.2×10^{13} watts. A portion of the core's thermal energy is transported toward the crust by mantle plumes; a form of convection consisting of upwellings of higher-temperature rock. These plumes can produce hotspots and flood basalts. More of the heat in the Earth is lost through plate tectonics, by mantle upwelling associated with mid-ocean ridges. The final major mode of heat loss is through conduction through the lithosphere, the majority of which occurs in the oceans because the crust there is much thinner than that of the continents.

Tectonic Plates

Earth's main plates

Plate name	***Area 10^6 km^2***
African Plate[note 12]	78.0
Antarctic Plate	60.9
Indo-Australian Plate	47.2
Eurasian Plate	67.8
North American Plate	75.9
South American Plate	43.6
Pacific Plate	103.3

The mechanically rigid outer layer of the Earth, the lithosphere, is broken into pieces called tectonic plates. These plates are rigid segments that move in relation to one another at one of three types of plate boundaries: Convergent boundaries, at which two plates come together, Divergent boundaries, at which two plates are pulled apart, and Transform boundaries, in which two plates slide past one another laterally. Earthquakes, volcanic activity, mountain-building, and oceanic trench formation can occur along these plate boundaries. The tectonic plates ride on top of the asthenosphere, the solid but less-viscous part of the upper mantle that can flow and move along with the plates, and their motion is strongly coupled with convection patterns inside the Earth's mantle.

As the tectonic plates migrate across the planet, the ocean floor is subducted under the leading edges of the plates at convergent boundaries. At the same time, the upwelling of mantle material at divergent boundaries creates mid-ocean ridges. The combination of these processes continually recycles the oceanic crust back into the mantle. Because of this recycling, most of the ocean floor is less than 100 million years in age. The oldest oceanic crust is located in the Western Pacific, and has an estimated age of about 200 million years. By comparison, the oldest dated continental crust is 4030 million years old. Other notable plates include the Indian Plate, the Arabian Plate, the Caribbean Plate, the Nazca Plate off the west coast of South America and the Scotia Plate in the southern Atlantic Ocean. The Australian Plate fused with the Indian Plate between 50 and 55 million years ago. The fastest-moving plates are the oceanic plates, with the Cocos Plate advancing at a rate of 75 mm/yr and the Pacific Plate moving 52–69 mm/yr. At the other extreme, the slowest-moving plate is the Eurasian Plate, progressing at a typical rate of about 21 mm/yr.

Surface

The Earth's terrain varies greatly from place to place. About 70.8% of the surface is covered by water, with much of the continental shelf below sea level. The submerged surface has mountainous features, including a globe-spanning mid-ocean ridge system, as well as undersea volcanoes, oceanic trenches, submarine canyons, oceanic plateaus and abyssal plains. The remaining 29.2% not covered by water consists of mountains, deserts, plains, plateaus, and other geomorphologies. The planetary surface undergoes reshaping over geological time periods because of tectonics and erosion. The surface features built up or deformed through plate tectonics are subject to steady weathering from precipitation, thermal cycles, and chemical effects. Glaciation, coastal erosion, the build-up of coral reefs, and large meteorite impacts also act to reshape the landscape.

The continental crust consists of lower density material such as the igneous rocks granite and andesite. Less common is basalt, a denser volcanic rock that is the primary constituent of the ocean floors. Sedimentary rock is formed from the accumulation of sediment that becomes compacted together. Nearly 75% of the continental surfaces are covered by sedimentary rocks, although they form only about 5% of the crust. The third form of rock material found on Earth is metamorphic rock, which is created from the transformation of pre-existing rock types through high pressures, high temperatures, or both. The most abundant silicate minerals on the Earth's surface include quartz, the feldspars, amphibole, mica, pyroxene and olivine. Common carbonate minerals include calcite (found in limestone) and dolomite. The pedosphere is the outermost layer of the Earth that is composed of soil and subject to soil formation processes. It exists at the interface of the lithosphere, atmosphere, hydrosphere and biosphere. Currently the total arable land is 13.31% of the land surface, with only 4.71% supporting permanent crops. Close to 40% of the Earth's land surface is presently used for cropland and pasture, or an estimated 1.3×10^7 km^2 of cropland and 3.4×10^7 km^2 of pastureland.

The elevation of the land surface of the Earth varies from the low point of “418 m at the Dead Sea, to a 2005-estimated maximum altitude of 8,848 m at the top of Mount Everest. The mean height of land above sea level is 840 m.

HYDROSPHERE

The abundance of water on Earth's surface is a unique feature that distinguishes the "Blue Planet" from others in the Solar System. The Earth's hydrosphere consists chiefly of the oceans, but technically includes all water surfaces in the world, including inland seas, lakes, rivers, and underground waters down to a depth of 2,000 m. The deepest underwater location is Challenger Deep of the Mariana Trench in the Pacific Ocean with a depth of “10,911.4 m. The mass of the oceans is approximately 1.35×10^{18} metric tons, or about 1/4400 of the total mass of the Earth. The oceans cover an area of 361.8×10^6 km^2 with a mean depth of 3,682 m, resulting in an estimated volume of 1.332×10^9 km^3. If all the land on Earth were spread evenly, water would rise to an altitude of more than 2.7 km. About 97.5% of the water is saline, while the remaining 2.5% is fresh water. Most fresh water, about 68.7%, is currently ice.

The average salinity of the Earth's oceans is about 35 grams of salt per kilogram of sea water (35 ‰). Most of this salt was released from volcanic activity or extracted from cool, igneous rocks. The oceans are also a reservoir of dissolved atmospheric gases, which are essential for the survival of many aquatic life forms. Sea water has an important influence on the world's climate, with the oceans acting as a large heat reservoir. Shifts in the oceanic temperature distribution can cause significant weather shifts, such as the El Niño-Southern Oscillation.

Atmosphere

The atmospheric pressure on the surface of the Earth averages 101.325 kPa, with a scale height of about 8.5 km. It is 78% nitrogen and 21% oxygen, with trace amounts of water vapour, carbon dioxide and other gaseous molecules. The height of the troposphere varies with latitude, ranging between 8 km at the poles to 17 km at the equator, with some variation resulting from weather and seasonal factors.

Earth's biosphere has significantly altered its atmosphere. Oxygenic photosynthesis evolved 2.7 billion years ago, forming the primarily nitrogen-oxygen atmosphere of today. This change enabled the proliferation of aerobic organisms as well as the formation of the ozone layer which blocks ultraviolet solar radiation, permitting life on land. Other atmospheric functions important to life on Earth include transporting water vapour, providing useful gases, causing small meteors to burn up before they strike the surface, and moderating temperature. This last phenomenon is known as the greenhouse effect: trace molecules within the atmosphere serve to capture thermal energy emitted from the ground, thereby raising the average temperature. Carbon dioxide, water vapour, methane and ozone are the primary greenhouse gases in the Earth's atmosphere. Without this heat-retention effect, the average surface temperature would be "18 °C and life would likely not exist.

Weather and Climate

The Earth's atmosphere has no definite boundary, slowly becoming thinner and fading into outer space. Three-quarters of the atmosphere's mass is contained within the first 11 km of the planet's surface. This lowest layer is called the troposphere. Energy from the Sun heats this layer, and the surface below, causing expansion of the air. This lower density air then rises, and is replaced by cooler, higher density air. The result is atmospheric circulation that drives the weather and climate through redistribution of heat energy. The primary atmospheric circulation bands consist of the trade winds in the equatorial region below 30° latitude and the westerlies in the mid-latitudes between 30° and 60°. Ocean currents are also important factors in determining climate, particularly the thermohaline circulation that distributes heat energy from the equatorial oceans to the polar regions.

Water vapour generated through surface evaporation is transported by circulatory patterns in the atmosphere. When atmospheric conditions permit an uplift of warm, humid air, this water condenses and settles to the surface as precipitation. Most of the water is then transported to lower elevations by river systems and usually returned to the oceans or deposited into lakes. This water cycle is a vital mechanism for supporting life on land, and is a primary factor in the erosion of surface features over geological periods. Precipitation patterns vary widely, ranging from several meters of water per year to less than a millimeter. Atmospheric circulation, topological features and temperature differences determine the average precipitation that falls in each region. The amount of solar energy reaching the Earth's decreases with increasing latitude. At higher latitudes the sunlight reaches the surface at a lower angles and it must pass through thicker columns of the atmosphere. As a result, the mean annual air temperature at sea level decreases by about 0.4°C per degree of latitude away from the equator. The Earth can be sub-divided into specific latitudinal belts of approximately homogeneous climate. Ranging from the equator to the polar regions, these are the tropical (or equatorial), subtropical, temperate and polar climates. Climate can also be classified based on the temperature and precipitation, with the climate regions characterized by fairly uniform air masses. The commonly used Köppen climate classification system (as modified by Wladimir Köppen's student Rudolph Geiger) has five broad groups (humid tropics, arid, humid middle latitudes, continental and cold polar), which are further divided into more specific subtypes.

Upper Atmosphere

Above the troposphere, the atmosphere is usually divided into the stratosphere, mesosphere, and thermosphere. Each layer has a different lapse rate, defining the rate of change in temperature with height. Beyond these, the exosphere thins out into the magnetosphere, where the Earth's magnetic fields interact with the solar wind. Within the stratosphere is the ozone layer, a component that partially shields the surface from ultraviolet light and thus is important for life on Earth. The Kármán line, defined as 100 km above the Earth's surface, is a working definition for the boundary between atmosphere and space.

Thermal energy causes some of the molecules at the outer edge of the Earth's atmosphere have their velocity increased to the point where they can escape from the planet's gravity. This results in a slow but steady leakage of the atmosphere into space. Because unfixed hydrogen has a low molecular weight, it can achieve escape velocity more readily and it leaks into outer space at a greater rate than other gasses. The leakage of hydrogen into space contributes to the pushing of the Earth from an initially reducing state to its current oxidizing one. Photosynthesis provided a source of free oxygen, but the loss of reducing agents such as hydrogen is believed to have been a necessary precondition for the widespread accumulation of oxygen in the atmosphere. Hence the ability of hydrogen to escape from the Earth's atmosphere may have influenced the nature of life that developed on the planet. In the current, oxygen-rich atmosphere most hydrogen is converted into water before it has an opportunity to escape. Instead, most of the hydrogen loss comes from the destruction of methane in the upper atmosphere.

Magnetic Field

The Earth's magnetic field is shaped roughly as a magnetic dipole, with the poles currently located proximate to the planet's geographic poles. According to dynamo theory, the field is generated within the molten outer core region where heat creates convection motions of conducting materials, generating electric currents. These in turn produce the Earth's magnetic field. The convection movements in the core are chaotic; the magnetic poles drift and periodically change alignment. This results in field reversals at irregular intervals averaging a few times every million years. The most recent reversal occurred approximately 700,000 years ago. The field forms the magnetosphere, which deflects particles in the solar wind. The sunward edge of the bow shock is located at about 13 times the radius of the Earth. The collision between the magnetic field and the solar wind forms the Van Allen radiation belts, a pair of concentric, torus-shaped regions of energetic charged particles. When the plasma enters the Earth's atmosphere at the magnetic poles, it forms the aurora.

ORBIT AND ROTATION

Rotation

Earth's rotation period relative to the Sun—its mean solar day—is 86,400 seconds of mean solar time (86,400.0025 SI seconds). As the Earth's solar day is now slightly longer than it was during the 19th century because of tidal acceleration, each day varies between 0 and 2 SI ms longer. Earth's rotation period relative to the fixed stars, called its *stellar day* by the International Earth Rotation and Reference Systems Service (IERS), is 86164.098903691 seconds of mean solar time (UT1), or $23^h\ 56^m\ 4.098903691^s$. Earth's rotation period relative to the precessing or moving mean vernal equinox, misnamed its *sidereal day*, is 86164.09053083288 seconds of mean solar time (UT1) ($23^h\ 56^m\ 4.09053083288^s$). Thus the sidereal day is shorter than the stellar day by about 8.4 ms. The length of the mean solar day in SI seconds is available from the IERS for the periods 1623–2005 and 1962–2005.

Apart from meteors within the atmosphere and low-orbiting satellites, the main apparent motion of celestial bodies in the Earth's sky is to the west at a rate of 15°/h = 15'/min. For bodies near the celestial equator, this is equivalent to an apparent diameter of the Sun or Moon every two minutes; from the planet's surface, the apparent sizes of the Sun and the Moon are approximately the same.

Orbit

Earth orbits the Sun at an average distance of about 150 million kilometres every 365.2564 mean solar days, or one sidereal year. From Earth, this gives an apparent movement of the Sun eastward with respect to the stars at a rate of about 1°/day, or a Sun or Moon diameter every 12 hours. Because of this motion, on average it takes 24 hours—a solar day—for

Earth to complete a full rotation about its axis so that the Sun returns to the meridian. The orbital speed of the Earth averages about 30 km/s (108,000 km/h), which is fast enough to cover the planet's diameter (about 12,600 km) in seven minutes, and the distance to the Moon (384,000 km) in four hours.

The Moon revolves with the Earth around a common barycenter every 27.32 days relative to the background stars. When combined with the Earth–Moon system's common revolution around the Sun, the period of the synodic month, from new moon to new moon, is 29.53 days. Viewed from the celestial north pole, the motion of Earth, the Moon and their axial rotations are all counter-clockwise. Viewed from a vantage point above the north poles of both the Sun and the Earth, the Earth appears to revolve in a counterclockwise direction about the Sun. The orbital and axial planes are not precisely aligned: Earth's axis is tilted some 23.5 degrees from the perpendicular to the Earth–Sun plane, and the Earth–Moon plane is tilted about 5 degrees against the Earth-Sun plane. Without this tilt, there would be an eclipse every two weeks, alternating between lunar eclipses and solar eclipses. The Hill sphere, or gravitational sphere of influence, of the Earth is about 1.5 Gm (or 1,500,000 kilometres) in radius. This is maximum distance at which the Earth's gravitational influence is stronger than the more distant Sun and planets. Objects must orbit the Earth within this radius, or they can become unbound by the gravitational perturbation of the Sun. Earth, along with the Solar System, is situated in the Milky Way galaxy, orbiting about 28,000 light years from the centre of the galaxy. It is currently about 20 light years above the galaxy's equatorial plane in the Orion spiral arm.

Axial Tilt and Seasons

Because of the axial tilt of the Earth, the amount of sunlight reaching any given point on the surface varies over the course of the year. This results in seasonal change in climate, with summer in the northern hemisphere occurring when the North Pole is pointing toward the Sun, and winter taking place when the pole is pointed away. During the summer, the day lasts longer and the Sun climbs higher in the sky. In winter, the climate becomes generally cooler and the days shorter. Above the Arctic Circle, an extreme case is reached where there is no daylight at all for part of the year—a polar night. In the southern hemisphere the situation is exactly reversed, with the South Pole oriented opposite the direction of the North Pole.

By astronomical convention, the four seasons are determined by the solstices—the point in the orbit of maximum axial tilt toward or away from the Sun—and the equinoxes, when the direction of the tilt and the direction to the Sun are perpendicular. In the northern hemisphere, Winter Solstice occurs on about December 21, Summer Solstice is near June 21, Spring Equinox is around March 20 and Autumnal Equinox is about September 23. In the Southern hemisphere, the situation is reversed, with the Summer and Winter Solstices exchanged and the Spring and Autumnal Equinox dates switched.

The angle of the Earth's tilt is relatively stable over long periods of time. However, the tilt does undergo nutation; a slight, irregular motion with a main period of 18.6 years. The orientation (rather than the angle) of the Earth's axis also changes over time, precessing around in a complete circle over each 25,800 year cycle; this precession is the reason for the difference between a sidereal year and a tropical year. Both of these motions are caused by the varying attraction of the Sun and Moon on the Earth's equatorial bulge. From the perspective of the Earth, the poles also migrate a few meters across the surface. This polar motion has multiple, cyclical components, which collectively are termed quasiperiodic motion. In addition to an annual component to this motion, there is a 14-month cycle called the Chandler wobble. The rotational velocity of the Earth also varies in a phenomenon known as length of day variation.

In modern times, Earth's perihelion occurs around January 3, and the aphelion around July 4. However, these dates change over time due to precession and other orbital factors, which follow cyclical patterns known as Milankovitch cycles. The changing Earth-Sun distance results in an increase of about 6.9% in solar energy reaching the Earth at perihelion relative to aphelion. Since the southern hemisphere is tilted toward the Sun at about the same time that the Earth reaches the closest approach to the Sun, the southern hemisphere receives slightly more energy from the Sun than does the northern over the course of a year. However, this effect is much less significant than the total energy change due to the axial tilt, and most of the excess energy is absorbed by the higher proportion of water in the southern hemisphere.

Moon

Characteristics

Diameter	***3,474.8 km***
Mass	7.349×10^{22} kg
Semi-major axis	384,400 km
Orbital period	27 d 7 h 43.7 m

The Moon is a relatively large, terrestrial, planet-like satellite, with a diameter about one-quarter of the Earth's. It is the largest moon in the Solar System relative to the size of its planet, although Charon is larger relative to the dwarf planet Pluto. The natural satellites orbiting other planets are called "moons" after Earth's Moon. The gravitational attraction between the Earth and Moon causes tides on Earth. The same effect on the Moon has led to its tidal locking: its rotation period is the same as the time it takes to orbit the Earth. As a result, it always presents the same face to the planet. As the Moon orbits Earth, different parts of its face are illuminated by the Sun, leading to the lunar phases; the dark part of the face is separated from the light part by the solar terminator.

Because of their tidal interaction, the Moon recedes from Earth at the rate of approximately 38 mm a year. Over millions of years, these tiny modifications—and the lengthening of Earth's day by about 23 µs a year—add up to significant changes. During the Devonian period, for example, (approximately 410 million years ago) there were 400 days in a year, with each day lasting 21.8 hours.

The Moon may have dramatically affected the development of life by moderating the planet's climate. Paleontological evidence and computer simulations show that Earth's axial tilt is stabilized by tidal interactions with the Moon. Some theorists believe that without this stabilization against the torques applied by the Sun and planets to the Earth's equatorial bulge, the rotational axis might be chaotically unstable, exhibiting chaotic changes over millions of years, as appears to be the case for Mars. If Earth's axis of rotation were to approach the plane of the ecliptic, extremely severe weather could result from the resulting extreme seasonal differences. One pole would be pointed directly toward the Sun during *summer* and directly away during *winter*. Planetary scientists who have studied the effect claim that this might kill all large animal and higher plant life. However, this is a controversial subject, and further studies of Mars—whose rotation period and axial tilt are similar to those of Earth, but which lacks a large moon or liquid core—may settle the matter. Viewed from Earth, the Moon is just far enough away to have very nearly the same apparent-sized disk as the Sun. The angular size (or solid angle) of these two bodies match because, although the Sun's diameter is about 400 times as large as the Moon's, it is also 400 times more distant. This allows total and annular solar eclipses to occur on Earth.

The most widely accepted theory of the Moon's origin, the giant impact theory, states that it formed from the collision of a Mars-size protoplanet called Theia with the early Earth. This hypothesis explains (among other things) the Moon's relative lack of iron and volatile elements, and the fact that its composition is nearly identical to that of the Earth's crust.

HABITABILITY

A planet that can sustain life is termed habitable, even if life did not originate there. The Earth provides the (currently understood) requisite conditions of liquid water, an environment where complex organic molecules can assemble, and sufficient energy to sustain metabolism. The distance of the Earth from the Sun, as well as its orbital eccentricity, rate of rotation, axial tilt, geological history, sustaining atmosphere and protective magnetic field all contribute to the conditions believed necessary to originate and sustain life on this planet.

Biosphere

The planet's life forms are sometimes said to form a "biosphere". This biosphere is generally believed to have begun evolving about 3.5 billion years ago. Earth is the only place where life is known to exist. The biosphere is divided into a number of biomes, inhabited by broadly similar plants and animals. On land, biomes are separated primarily by differences in latitude, height above sea level and humidity. Terrestrial biomes lying within the Arctic or Antarctic Circles, at high altitudes or in extremely arid areas are relatively barren of plant and animal life; species diversity reaches a peak in humid lowlands at equatorial latitudes.

Natural Resources and Land Use

The Earth provides resources that are exploitable by humans for useful purposes. Some of these are non-renewable resources, such as mineral fuels, that are difficult to replenish on a short time scale. Large deposits of fossil fuels are obtained from the Earth's crust, consisting of coal, petroleum, natural gas and methane clathrate.

These deposits are used by humans both for energy production and as feedstock for chemical production. Mineral ore bodies have also been formed in Earth's crust through a process of Ore genesis, resulting from actions of erosion and plate tectonics. These bodies form concentrated sources for many metals and other useful elements.

The Earth's biosphere produces many useful biological products for humans, including (but far from limited to) food, wood, pharmaceuticals, oxygen, and the recycling of many organic wastes. The land-based ecosystem depends upon topsoil and fresh water, and the oceanic ecosystem depends upon dissolved nutrients washed down from the land. Humans also live on the land by using building materials to construct shelters. In 1993, human use of land is approximately:

Land use	*Arable land*	*Permanent crops*	*Permanent pastures*	*Forests & woodland*	*Urban areas*	*Other*
Percentage	13.13%	4.71%	26%	32%	1.5%	30%

The estimated amount of irrigated land in 1993 was 2,481,250 km^2.

Natural and Environmental Hazards

Large areas of the Earth's surface are subject to extreme weather such as tropical cyclones, hurricanes, or typhoons that dominate life in those areas. Many places are subject to earthquakes, landslides, tsunamis, volcanic eruptions, tornadoes, sinkholes, blizzards, floods, droughts, and other calamities and disasters. Many localized areas are subject to human-made pollution of the air and water, acid rain and toxic substances, loss of vegetation (overgrazing, deforestation, desertification), loss of wildlife, species extinction, soil degradation, soil depletion, erosion, and introduction of invasive species.

According to the United Nations, a scientific consensus exists linking human activities to global warming due to industrial carbon dioxide emissions. This is predicted to produce changes such as the melting of glaciers and ice sheets, more extreme temperature ranges, significant changes in weather and a global rise in average sea levels.

Human Geography

Earth has approximately 6,803,000,000 human inhabitants as of December 12, 2009. Projections indicate that the world's human population will reach seven billion in 2013 and 9.2 billion in 2050. Most of the growth is expected to take place in developing nations. Human population density varies widely around the world, but a majority live in Asia. By 2020, 60% of the world's population is expected to be living in urban, rather than rural, areas. It is estimated that only one-eighth of the surface of the Earth is suitable for humans to live on—three-quarters is covered by oceans, and half of the land area is either desert (14%), high mountains (27%), or other less suitable terrain. The northernmost permanent settlement in the world is Alert, on Ellesmere Island in Nunavut, Canada. (82°282 N) The southernmost is the Amundsen-Scott South Pole Station, in Antarctica, almost exactly at the South Pole. (90°S)

Independent sovereign nations claim the planet's entire land surface, except for some parts of Antarctica and the odd unclaimed area of Bir Tawil between Egypt and Sudan. As of 2007 there are 201 sovereign states, including the 192 United Nations member states. In addition, there are 59 dependent territories, and a number of autonomous areas, territories under dispute and other entities. Historically, Earth has never had a sovereign government with authority over the entire globe, although a number of nation-states have striven for world domination and failed. The United Nations is a worldwide intergovernmental organization that was created with the goal of intervening in the disputes between nations, thereby avoiding armed conflict.

It is not, however, a world government. The U.N. serves primarily as a forum for international diplomacy and international law. When the consensus of the membership permits, it provides a mechanism for armed intervention.

The first human to orbit the Earth was Yuri Gagarin on April 12, 1961. In total, about 400 people visited outer space and reached Earth orbit as of 2004, and, of these, twelve have walked on the Moon. Normally the only humans in space are those on the International Space Station. The station's crew, currently six people, is usually replaced every six months. The furthest humans have travelled from Earth is 400,171 km, achieved during the 1970 Apollo 13 mission.

CULTURAL VIEWPOINT

The name "Earth" derives from the Anglo-Saxon word *erda*, which means ground or soil, and is related to the German word *erde*. It became *eorthe* later, and then *erthe* in Middle English. The standard astronomical symbol of the Earth consists of a cross circumscribed by a circle.

Unlike the rest of the planets in the Solar System, mankind did not perceive the Earth as a planet until the 16th century. Earth has often been personified as a deity, in particular a goddess. In many cultures the mother goddess is also portrayed as a fertility deity. Creation myths in many religions recall a story involving the creation of the Earth by a supernatural deity or deities. A variety of religious groups, often associated with fundamentalist branches of Protestantism or Islam, assert that their interpretations of these creation myths in sacred texts are literal truth and should be considered alongside or replace conventional scientific accounts of the formation of the Earth and the origin and development of life. Such assertions are opposed by the scientific community and by other religious groups. A prominent example is the creation-evolution controversy.

In the past there were varying levels of belief in a flat Earth, but this was displaced by the concept of a spherical

Earth due to observation and circumnavigation. The human perspective regarding the Earth has changed following the advent of spaceflight, and the biosphere is now widely viewed from a globally integrated perspective. This is reflected in a growing environmental movement that is concerned about humankind's effects on the planet.

STRUCTURE OF THE EARTH

The interior structure of the Earth, similar to the outer, is layered. These layers can be defined by either their chemical or their rheological properties.

The Earth has an outer silicate solid crust, a highly viscous mantle, a liquid outer core that is much less viscous than the mantle, and a solid inner core. Scientific understanding of Earth's internal structure is based on observations of topography and bathymetry, observations of rock in outcrop, samples brought to the surface from greater depths by volcanic activity, analysis of the seismic waves that pass through the Earth, measurements of the gravity field of the Earth, and experiments with crystalline solids at pressures and temperatures characteristic of the Earth's deep interior.

Assumptions

The force exerted by Earth's gravity can be used to calculate its mass, and by estimating the volume of the planet, its average density can be calculated. Astronomers can also calculate Earth's mass from its orbit and effects on nearby planetary bodies. Observations of rocks, bodies of water and atmosphere allow estimation of the mass, volume and density of rocks to a certain depth, so the remaining mass must be in the deeper layers.

Structure

The structure of Earth can be defined in two ways: by mechanical properties such as rheology, or chemically. Mechanically, it can be divided into lithosphere, asthenosphere, mesosphere, outer core, and the inner core. The interior of the earth is divided into 5 important layers. Chemically, Earth can be divided into the crust, upper mantle, lower mantle, outer core, and inner core. The geologic component layers of Earth are at the following depths below the surface:

Depth		
Kilometres	***Miles***	***Layer***
0–60	0–37	Lithosphere (locally varies between 5 and 200 km)
0–35	0–22	... Crust (locally varies between 5 and 70 km)
35–60	22–37	... Uppermost part of mantle
35–2,890	22–1,790	Mantle
100–200	62–125	... Asthenosphere
35–660	22–410	... Upper mantle
660–2,890	410–1,790	... Lower mantle
2,890–5,150	1,790–3,160	Outer core
5,150–6,360	3,160–3,954	Inner core

The layering of Earth has been inferred indirectly using the time of travel of refracted and reflected seismic waves created by earthquakes. The core does not allow shear waves to pass through it, while the speed of travel (seismic velocity) is different in other layers. The changes in seismic velocity between different layers causes refraction owing to Snell's law. Reflections are caused by a large increase in seismic velocity and are similar to light reflecting from a mirror.

Core

The average density of Earth is 5,515 kg/m^3. Since the average density of surface material is only around 3,000 kg/m^3, we must conclude that denser materials exist within Earth's core. Further evidence for the high density core comes from the study of seismology.

Seismic measurements show that the core is divided into two parts, a solid inner core with a radius of ~1,220 km and a liquid outer core extending beyond it to a radius of ~3,400 km. The solid inner core was discovered in 1936 by Inge Lehmann and is generally believed to be composed primarily of iron and some nickel. In early stages of Earth's formation about 4.5 billion (4.5×10^9) years ago, melting would have caused denser substances to sink toward the centre in a process called planetary differentiation, while less-dense materials would have migrated to the crust. The core is thus believed to largely be composed of iron (80%), along with nickel and one or more light elements, whereas other dense elements, such as lead and uranium, either are too rare to be significant or tend to bind to lighter elements and thus remain in the crust. Some have argued that the inner core may be in the form of a single iron crystal.

The liquid outer core surrounds the inner core and is believed to be composed of iron mixed with nickel and trace amounts of lighter elements. Recent speculation suggests that the innermost part of the core is enriched in gold, platinum and other iron-loving (siderophile) elements.

The matter that comprises Earth is connected in fundamental ways to matter of certain chondrite meteorites, and to matter of outer portion of the Sun. There is good reason to believe that Earth is, in the main, like a chondrite meteorite. Beginning as early as 1940, scientists, including Francis Birch, built geophysics upon the premise that Earth is like ordinary chondrites, the most common type of meteorite observed impacting Earth, while totally ignoring another, albeit less abundant type, called enstatite chondrites. The principal difference between the two meteorite types is that enstatite chondrites formed under circumstances of extremely limited available oxygen, leading to certain normally oxyphile elements existing either partially or wholly in the alloy portion that corresponds to the core of Earth.

Dynamo theory suggests that convection in the outer core, combined with the Coriolis effect, gives rise to Earth's magnetic field. The solid inner core is too hot to hold a permanent magnetic field but probably acts to stabilize the magnetic field generated by the liquid outer core. Recent evidence has suggested that the inner core of Earth may rotate slightly faster than the rest of the planet. In August 2005 a team of geophysicists announced in the journal *Science* that, according to their estimates, Earth's inner core rotates approximately 0.3 to 0.5 degrees per year relative to the rotation of the surface.

The current scientific explanation for the Earth's temperature gradient is a combination of heat left over from the planet's initial formation, decay of radioactive elements, and freezing of the inner core.

Mantle

Earth's mantle extends to a depth of 2,890 km, making it the thickest layer of the Earth. The pressure, at the bottom of the mantle, is ~140 GPa (1.4 Matm). The mantle is composed of silicate rocks that are rich in iron and magnesium relative to the overlying crust. Although solid, the high temperatures within the mantle cause the silicate material to be sufficiently ductile that it can flow on very long timescales. Convection of the mantle is expressed at the surface through the motions of tectonic plates. The melting point and viscosity of a substance depends on the pressure it is under. As there is intense and increasing pressure as one travels deeper into the mantle, the lower part of the mantle flows less easily than does the upper mantle (chemical changes within the mantle may also be important). The viscosity of the mantle ranges between 10^{21} and 10^{24} Pa·s, depending on depth. In comparison, the viscosity of water is approximately 10^{-3} Pa·s and that of pitch is 10^{7} Pa·s.

Crust

The crust ranges from 5–70 km in depth and is the outermost layer. The thin parts are the oceanic crust, which underlie the ocean basins(5–10 km) and are composed of dense (mafic) iron magnesium silicate rocks, like basalt. The thicker crust is continental crust, which is not less dense and composed of (felsic) sodium potassium aluminium silicate rocks, like granite. The rocks of the crust fall into two major categories-sial and sima (Suess,1831–1914). As the main mineral constituents of the continental mass are silica and alumina, it is thus called sial (si-silica, 65–75% and al-alumina). The oceanic crust mainly consists of silica and magnesium; it is therefore called sima (si-silica and ma-magnesium). It is estimated that sima starts about 11 km below the Conrad discontinuity, a second order discontinuity. The uppermost mantle together with the crust constitutes the lithosphere. The crust-mantle boundary occurs as two physically different events. First, there is a discontinuity in the seismic velocity, which is known as the Mohorovièiæ discontinuity or Moho. The cause of the Moho is thought to be a change in rock composition from rocks containing plagioclase feldspar (above) to rocks that contain no feldspars (below). Second, in oceanic crust, there is a chemical discontinuity between ultramafic cumulates and tectonized harzburgites, which has been observed from deep parts of the oceanic crust that have been obducted onto the continental crust and preserved as ophiolite sequences.

Many rocks now making up Earth's crust formed less than 100 million (1×10^8) years ago; however the oldest known mineral grains are 4.4 billion (4.4×10^9) years old, indicating that Earth has had a solid crust for at least that long.

Historical Development of Alternative Conceptions

In 1692 Edmund Halley (in a paper printed in *Philosophical Transactions of Royal Society of London*) put forth the idea of Earth consisting of a hollow shell about 500 miles thick, with two inner concentric shells around an innermost core, corresponding to the diameters of the planets Venus, Mars, and Mercury respectively. Halley's construct was a method of accounting for the (flawed) values of the relative density of Earth and the Moon that had been given by Sir Isaac Newton, in *Principia* (1687). "Sir Isaac Newton has demonstrated the Moon to be more solid than our Earth, as 9 to 5," Halley remarked; "why may we not then suppose four ninths of our globe to be cavity?"

3

The Atmosphere

ATMOSPHERE OF EARTH

The atmosphere of Earth is a layer of gases surrounding the planet Earth that is retained by Earth's gravity. The atmosphere protects life on Earth by absorbing ultraviolet solar radiation, warming the surface through heat retention (greenhouse effect), and reducing temperature extremes between day and night.

Atmospheric stratification describes the structure of the atmosphere, dividing it into distinct layers, each with specific characteristics such as temperature or composition. The atmosphere has a mass of about 5×10^{18} kg, three quarters of which is within about 11 km (6.8 mi; 36,000 ft) of the surface. The atmosphere becomes thinner and thinner with increasing altitude, with no definite boundary between the atmosphere and outer space. An altitude of 120 km (75 mi) is where atmospheric effects become noticeable during atmospheric reentry of spacecraft. The Kármán line, at 100 km (62 mi), also is often regarded as the boundary between atmosphere and outer space.

Air is the name given to atmosphere used in breathing and photosynthesis. Dry air contains roughly (by volume) 78.09% nitrogen, 20.95% oxygen, 0.93% argon, 0.039% carbon dioxide, and small amounts of other gases. Air also contains a variable amount of water vapour, on average around 1%. While air content and atmospheric pressure varies at different layers, air suitable for the survival of terrestrial plants and terrestrial animals is currently known only to be found in Earth's troposphere and artificial atmospheres.

Composition

Air is mainly composed of nitrogen, oxygen, and argon, which together constitute the major gases of the atmosphere. The remaining gases are often referred to as trace gases, among which are the greenhouse gases such as water vapour, carbon dioxide, methane, nitrous oxide, and ozone. Filtered air includes trace amounts of many other chemical compounds. Many natural substances may be present in tiny amounts in an unfiltered air sample, including dust, pollen and spores, sea spray, and volcanic ash. Various industrial pollutants also may be present, such as chlorine (elementary or in compounds), fluorine compounds, elemental mercury, and sulfur compounds such as sulfur dioxide [SO_2].

Composition of dry atmosphere

Gas	*Volume*
Nitrogen (N_2)	780,840 ppmv (78.084%)
Oxygen (O_2)	209,460 ppmv (20.946%)
Argon (Ar)	9,340 ppmv (0.9340%)
Carbon dioxide (CO_2)	390 ppmv (0.039%)
Neon (Ne)	18.18 ppmv (0.001818%)
Helium (He)	5.24 ppmv (0.000524%)
Methane (CH_4)	1.79 ppmv (0.000179%)
Krypton (Kr)	1.14 ppmv (0.000114%)
Hydrogen (H_2)	0.55 ppmv (0.000055%)
Nitrous oxide (N_2O)	0.3 ppmv (0.00003%)
Carbon monoxide (CO)	0.1 ppmv (0.00001%)
Xenon (Xe)	0.09 ppmv (9×10^{-6}%) (0.000009%)
Ozone (O_3)	0.0 to 0.07 ppmv (0 to 7×10^{-6}%)
Nitrogen dioxide (NO_2)	0.02 ppmv (2×10^{-6}%) (0.000002%)
Iodine (I_2)	0.01 ppmv (1×10^{-6}%) (0.000001%)
Ammonia (NH_3)	trace
Not included in above dry atmosphere:	
Water vapour (H_2O)	~0.40% over full atmosphere, typically 1%-4% at surface

STRUCTURE OF THE ATMOSPHERE

Principal Layers

Earth's atmosphere can be divided into five main layers. These layers are mainly determined by whether temperature increases or decreases with altitude. From highest to lowest, these layers are:

Exosphere

The outermost layer of Earth's atmosphere extends from the exobase upward. It is mainly composed of hydrogen and helium.

The particles are so far apart that they can travel hundreds of kilometres without colliding with one another. Since the particles rarely collide, the atmosphere no longer behaves like a fluid. These free-moving particles follow ballistic trajectories and may migrate into and out of the magnetosphere or the solar wind.

Thermosphere

Temperature increases with height in the thermosphere from the mesopause up to the thermopause, then is constant with height. The temperature of this layer can rise to 1,500 °C (2,730 °F), though the gas molecules are so far apart that temperature in the usual sense is not well defined. The International Space Station orbits in this layer, between 320 and 380 km (200 and 240 mi). The top of the thermosphere is the bottom of the exosphere, called the exobase. Its height varies with solar activity and ranges from about 350–800 km (220–500 mi; 1,100,000–2,600,000 ft).

Mesosphere

The mesosphere extends from the stratopause to 80–85 km (50–53 mi; 260,000–280,000 ft). It is the layer where most meteors burn up upon entering the atmosphere. Temperature decreases with height in the mesosphere. The mesopause, the temperature minimum that marks the top of the mesosphere, is the coldest place on Earth and has an average temperature around "85 °C ("121 °F; 188.1 K). Due to the cold temperature of the mesophere, water vapour is frozen, forming ice clouds (or Noctilucent clouds). A type of lightning referred to as either sprites or ELVES, form many miles above thunderclouds in the troposphere.

Stratosphere

The stratosphere extends from the tropopause to about 51 km (32 mi; 170,000 ft). Temperature increases with height, which restricts turbulence and mixing. The stratopause, which is the boundary between the stratosphere and mesosphere, typically is at 50 to 55 km (31 to 34 mi; 160,000 to 180,000 ft). The pressure here is 1/1000 sea level.

Troposphere

The troposphere begins at the surface and extends to between 7 km (23,000 ft) at the poles and 17 km (56,000 ft) at the equator, with some variation due to weather.

The troposphere is mostly heated by transfer of energy from the surface, so on average the lowest part of the troposphere is warmest and temperature decreases with altitude. The troposphere contains roughly 80% of the mass of the atmosphere. The tropopause is the boundary between the troposphere and stratosphere.

Other Layers

Within the five principal layers determined by temperature are several layers determined by other properties.

- The ozone layer is contained within the stratosphere. In this layer ozone concentrations are about 2 to 8 parts per million, which is much higher than in the lower atmosphere but still very small compared to the main components of the atmosphere. It is mainly located in the lower portion of the stratosphere from about 15–35 km (9.3–22 mi; 49,000–110,000 ft), though the thickness varies seasonally and geographically. About 90% of the ozone in our atmosphere is contained in the stratosphere.
- The ionosphere, the part of the atmosphere that is ionized by solar radiation, stretches from 50 to 1,000 km (31 to 620 mi; 160,000 to 3,300,000 ft) and typically overlaps both the exosphere and the thermosphere. It forms the inner edge of the magnetosphere. It has practical importance because it influences, for example,

radio propagation on the Earth. It is responsible for auroras.

- The homosphere and heterosphere are defined by whether the atmospheric gases are well mixed. In the homosphere the chemical composition of the atmosphere does not depend on molecular weight because the gases are mixed by turbulence. The homosphere includes the troposphere, stratosphere, and mesosphere. Above the *turbopause* at about 100 km (62 mi; 330,000 ft) (essentially corresponding to the mesopause), the composition varies with altitude. This is because the distance that particles can move without colliding with one another is large compared with the size of motions that cause mixing. This allows the gases to stratify by molecular weight, with the heavier ones such as oxygen and nitrogen present only near the bottom of the heterosphere. The upper part of the heterosphere is composed almost completely of hydrogen, the lightest element.
- The planetary boundary layer is the part of the troposphere that is nearest the Earth's surface and is directly affected by it, mainly through turbulent diffusion. During the day the planetary boundary layer usually is well-mixed, while at night it becomes stably stratified with weak or intermittent mixing. The depth of the planetary boundary layer ranges from as little as about 100 m on clear, calm nights to 3000 m or more during the afternoon in dry regions.

The average temperature of the atmosphere at the surface of Earth is 14 °C (57 °F; 287 K) or 15 °C (59 °F; 288 K), depending on the reference.

PHYSICAL PROPERTIES

Pressure and Thickness

The average atmospheric pressure at sea level is about 1 atmosphere (atm) = 101.3 kPa (kilopascals) = 14.7 psi (pounds per square inch) = 760 torr = 29.9 inches of mercury (symbol Hg). Total atmospheric mass is 5.1480×10^{18} kg (1.135×10^{19} lb), about 2.5% less than would be inferred naively from the average sea level pressure and the Earth's area of 51007.2 megahectares, this defect having been displaced by the Earth's mountainous terrain. Atmospheric pressure is the total weight of the air above unit area at the point where the pressure is measured. Thus air pressure varies with location and time, because the amount of air above the Earth's surface varies.

If atmospheric density were to remain constant with height the atmosphere would terminate abruptly at 8.50 km (27,900 ft). Instead, density decreases with height, dropping by 50% at an altitude of about 5.6 km (18,000 ft). As a result the pressure decrease is approximately exponential with height, so that pressure decreases by a factor of two approximately every 5.6 km (18,000 ft) and by a factor of $e = 2.718\ldots$ approximately every 7.64 km (25,100 ft), the latter being the average scale height of Earth's atmosphere below 70 km (43 mi; 230,000 ft). However, because of changes in temperature, average molecular weight, and gravity throughout the atmospheric column, the dependence of atmospheric pressure on altitude is modeled by separate equations for each of the layers listed above. Even in the exosphere, the atmosphere is still present. This can be seen by the effects of atmospheric drag on satellites.

In summary, the equations of pressure by altitude in the above references can be used directly to estimate atmospheric thickness. However, the following published data are given for reference:

- 50% of the atmosphere by mass is below an altitude of 5.6 km (18,000 ft).
- 90% of the atmosphere by mass is below an altitude of 16 km (52,000 ft). The common altitude of commercial airliners is about 10 km (33,000 ft) and Mt. Everest's summit is 8,848 m (29,029 ft) above sea level.
- 99.99997% of the atmosphere by mass is below 100 km (62 mi; 330,000 ft), although in the rarefied region above this there are auroras and other atmospheric effects. The highest X-15 plane flight in 1963 reached an altitude of 108.0 km (354,300 ft).

Density and Mass

The density of air at sea level is about 1.2 kg/m^3 (1.2 g/L). Density is not measured directly but is calculated from measurements of temperature, pressure and humidity using the equation of state for air (a form of the ideal gas law). Atmospheric density decreases as the altitude increases. This variation can be approximately modeled using the barometric formula. More sophisticated models are used to predict orbital decay of satellites.

The average mass of the atmosphere is about 5 quadrillion (5×10^{15}) tonnes or 1/1,200,000 the mass of Earth.

According to the National Centre for Atmospheric Research, "The total mean mass of the atmosphere is 5.1480×10^{18} kg with an annual range due to water vapour of 1.2 or 1.5×10^{15} kg depending on whether surface pressure or water vapour data are used; somewhat smaller than the previous estimate. The mean mass of water vapour is estimated as 1.27×10^{16} kg and the dry air mass as $5.1352 \pm 0.0003 \times 10^{18}$ kg."

Optical Properties

Solar radiation (or sunlight) is the energy the Earth receives from the Sun. The Earth also emits radiation back into space, but at longer wavelengths that we cannot see. Part of the incoming and emitted radiation is absorbed or reflected by the atmosphere.

Scattering

When light passes through our atmosphere, photons interact with it through *scattering*. If the light does not interact with the atmosphere, it is called *direct radiation* and is what you see if you were to look directly at the Sun.

Indirect radiation is light that has been scattered in the atmosphere. For example, on an overcast day when you cannot see your shadow there is no direct radiation reaching you, it has all been scattered. As another example, due to a phenomenon called Rayleigh scattering, shorter (blue) wavelengths scatter more easily than longer (red) wavelengths.

This is why the sky looks blue, you are seeing scattered blue light. This is also why sunsets are red. Because the Sun is close to the horizon, the Sun's rays pass through more atmosphere than normal to reach your eye. Much of the blue light has been scattered out, leaving the red light in a sunset.

Absorption

Different molecules absorb different wavelengths of radiation. For example, O_2 and O_3 absorb almost all wavelengths shorter than 300 nanometers. Water (H_2O) absorbs many wavelengths above 700 nm. When a molecule absorbs a photon, it increases the energy of the molecule. We can think of this as heating the atmosphere, but the atmosphere also cools by emitting radiation.

The combined absorption spectra of the gases in the atmosphere leave "windows" of low opacity, allowing the transmission of only certain bands of light. The optical window runs from around 300 nm (ultraviolet-C) up into the range humans can see, the visible spectrum (commonly called light), at roughly 400–700 nm and continues to the infrared to around 1100 nm. There are also infrared and radio windows that transmit some infrared and radio waves at longer wavelengths. For example, the radio window runs from about one centimetre to about eleven-meter waves.

Emission

Emission is the opposite of absorption, it is when an object emits radiation. Objects tend to emit amounts and wavelengths of radiation depending on their "black body" emission curves, therefore hotter objects tend to emit more radiation, with shorter wavelengths. Colder objects emit less radiation, with longer wavelengths. For example, the Sun is approximately 6,000 K (5,730 °C; 10,340 °F), its radiation peaks near 500 nm, and is visible to the human eye. The Earth is approximately 290 K (17 °C; 62 °F), so its radiation peaks near 10,000 nm, and is much too long to be visible to humans. Because of its temperature, the atmosphere emits infrared radiation. For example, on clear nights the Earth's surface cools down faster than on cloudy nights. This is because clouds (H_2O) are strong absorbers and emitters of infrared radiation. This is also why it becomes colder at night at higher elevations. The atmosphere acts as a "blanket" to limit the amount of radiation the Earth loses into space.

The *greenhouse effect* is directly related to this absorption and emission (or "blanket") effect. Some chemicals in the atmosphere absorb and emit infrared radiation, but do not interact with sunlight in the visible spectrum. Common examples of these chemicals are CO_2 and H_2O. If there are too much of these *greenhouse gases*, sunlight heats the Earth's surface, but the gases block the infrared radiation from exiting back to space. This imbalance causes the Earth to warm, and thus climate change.

Refractive Index

The refractive index of air is close to, but just greater than 1. Systematic variations in refractive index can lead to the bending of light rays over long optical paths. One example is that, under some circumstances, observers onboard ships can see other vessels just over the horizon because light is refracted in the same direction as the curvature of the Earth's surface. The refractive index of air depends on temperature, giving rise to refraction effects when the temperature gradient is large. An example of such effects is the mirage.

CIRCULATION

Atmospheric circulation is the large-scale movement of air, and the means (with ocean circulation) by which heat is distributed around the Earth. The large-scale structure of the atmospheric circulation varies from year to year, but the basic structure remains fairly constant as it is determined by the Earth's rotation rate and the difference in solar radiation between the equator and poles.

EVOLUTION OF EARTH'S ATMOSPHERE

Second Atmosphere

Water related sediments have been found dating from as early as 3.8 billion years ago. About 3.4 billion years ago, nitrogen was the major part of the then stable "second atmosphere." An influence of life has to be taken into account rather soon in the history of the atmosphere, since hints of early life forms are to be found as early as 3.5 billion years ago. The fact that this is not perfectly in line with the-compared to today 30% lower-solar radiance of the early Sun has been described as the "Faint young Sun paradox". The geological record however shows a continually relatively warm surface during the complete early temperature record of the Earth with the exception of one cold glacial phase about 2.4 billion years ago. In the late Archaean era an oxygen-containing atmosphere began to develop, apparently from photosynthesizing algae which have been found as stromatolite fossils from 2.7 billion years ago. The early basic carbon isotopy (isotope ratio proportions) is very much in line with what is found today, suggesting that the fundamental features of the carbon cycle were established as early as 4 billion years ago.

Third Atmosphere

The accretion of continents about 3.5 billion years ago added plate tectonics, constantly rearranging the continents and also shaping long-term climate evolution by allowing the transfer of carbon dioxide to large land-based carbonate storages. Free oxygen did not exist until about 1.7 billion years ago and this can be seen with the development of the red beds and the end of the banded iron formations. This signifies a shift from a reducing atmosphere to an oxidising atmosphere. O_2 showed major ups and downs until reaching a steady state of more than 15%. The following time span was the Phanerozoic era, during which oxygen-breathing metazoan life forms began to appear. Currently, anthropogenic greenhouse gases are increasing in the atmosphere. According to the Intergovernmental Panel on Climate Change, this increase is the main cause of global warming.

Air Pollution

Air pollution is the introduction of chemicals, particulate matter, or biological materials that cause harm or discomfort to organisms into the atmosphere. Stratospheric ozone depletion is believed to be caused by air pollution (chiefly from chlorofluorocarbons).

ATMOSPHERIC CHEMISTRY

Atmospheric chemistry is a branch of atmospheric science in which the chemistry of the Earth's atmosphere and that of other planets is studied. It is a multidisciplinary field of research and draws on environmental chemistry, physics, meteorology, computer modeling, oceanography, geology and volcanology and other disciplines. Research is increasingly connected with other areas of study such as climatology.

The composition and chemistry of the atmosphere is of importance for several reasons, but primarily because of the interactions between the atmosphere and living organisms. The composition of the Earth's atmosphere has been changed by human activity and some of these changes are harmful to human health, crops and ecosystems. Examples of problems which have been addressed by atmospheric chemistry include acid rain, ozone depletion, photochemical smog and global warming. Atmospheric chemistry seeks to understand the causes of these problems, and by obtaining a theoretical understanding of them, allow possible solutions to be tested and the effects of changes in government policy evaluated.

Atmospheric Composition

Average Composition of Dry Atmosphere

Gas	per NASA
Nitrogen, N_2	78.084%
Oxygen, O_2	20.946%
Argon, Ar	0.934%
Minor Constituents (in ppm)	
Carbon Dioxide, CO_2	383
Neon, Ne	18.18
Helium, He	5.24
Methane, CH_4	1.7
Krypton, Kr	1.14
Hydrogen, H_2	0.55
Water	
Water vapour	Highly variable; typically makes up about 1%

Notes: the concentration of CO_2 and CH_4 vary by season and location. The mean molecular mass of air is 28.97 g/mol.

History

The ancient Greeks regarded air as one of the four elements, but the first scientific studies of atmospheric composition began in the 18th century. Chemists such as Joseph Priestley, Antoine Lavoisier and Henry Cavendish made the first measurements of the composition of the atmosphere.

In the late 19th and early 20th centuries interest shifted towards trace constituents with very small concentrations. One particularly important discovery for atmospheric chemistry was the discovery of ozone by Christian Friedrich Schoenbein in 1840.

In the 20th century atmospheric science moved on from studying the composition of air to a consideration of how the concentrations of trace gases in the atmosphere have changed over time and the chemical processes which create and destroy compounds in the air. Two particularly important examples of this were the explanation of how the ozone layer is created and maintained by Sydney Chapman and Gordon Dobson, and the explanation of Photochemical smog by Arie Jan Haagen-Smit. Further studies on ozone issues led to the 1995 Nobel Prize in Chemistry award shared between Paul Crutzen, Mario Molina and Frank Sherwood Rowland.

In the 21st century the focus is now shifting again. Atmospheric chemistry is increasingly studied as one part of the Earth system. Instead of concentrating on atmospheric chemistry in isolation the focus is now on seeing it as one part of a single system with the rest of the atmosphere, biosphere and geosphere. An especially important driver for this is the links between chemistry and climate such as the effects of changing climate on the recovery of the ozone hole and vice versa but also interaction of the composition of the atmosphere with the oceans and terrestrial ecosystems.

Methodology

Observations, lab measurements and modeling are the three central elements in atmospheric chemistry. Progress in atmospheric chemistry is often driven by the interactions between these components and they form an integrated whole. For example observations may tell us that more of a chemical compound exists than previously thought possible. This will stimulate new modelling and laboratory studies which will increase our scientific understanding to a point where the observations can be explained.

Observation

Observations of atmospheric chemistry are essential to our

understanding. Routine observations of chemical composition tell us about changes in atmospheric composition over time. One important example of this is the Keeling Curve-a series of measurements from 1958 to today which show a steady rise in of the concentration of carbon dioxide. Observations of atmospheric chemistry are made in observatories such as that on Mauna Loa and on mobile platforms such as aircraft (e.g. the UK's Facility for Airborne Atmospheric Measurements), ships and balloons. Observations of atmospheric composition are increasingly made by satellites with important instruments such as GOME and MOPITT giving a global picture of air pollution and chemistry. Surface observations have the advantage that they provide long term records at high time resolution but are limited in the vertical and horizontal space they provide observations from. Some surface based instruments e.g. LIDAR can provide concentration profiles of chemical compounds and aerosol but are still restricted in the horizontal region they can cover. Many observations are available on line in Atmospheric Chemistry Observational Databases.

Lab Measurements

Measurements made in the laboratory are essential to our understanding of the sources and sinks of pollutants and naturally occurring compounds. Lab studies tell us which gases react with each other and how fast they react. Measurements of interest include reactions in the gas phase, on surfaces and in water. Also of high importance is photochemistry which quantifies how quickly molecules are split apart by sunlight and what the products are plus thermodynamic data such as Henry's law coefficients.

Modeling

In order to synthesise and test theoretical understanding of atmospheric chemistry, computer models (such as chemical transport models) are used. Numerical models solve the differential equations governing the concentrations of chemicals in the atmosphere. They can be very simple or very complicated. One common trade off in numerical models is between the number of chemical compounds and chemical reactions modelled versus the representation of transport and mixing in the atmosphere. For example, a box model might include hundreds or even thousands of chemical reactions but will only have a very crude representation of mixing in the atmosphere. In contrast, 3D models represent many of the physical processes of the atmosphere but due to constraints on computer resources will have far fewer chemical reactions and compounds. Models can be used to interpret observations, test understanding of chemical reactions and predict future concentrations of chemical compounds in the atmosphere. One important current trend is for atmospheric chemistry modules to become one part of earth system models in which the links between climate, atmospheric composition and the biosphere can be studied.

Some models are constructed by automatic code generators (e.g. Autochem or KPP). In this approach a set of constituents are chosen and the automatic code generator will then select the reactions involving those constituents from a set of reaction databases. Once the reactions have been chosen the ordinary differential equations (ODE) that describe their time evolution can be automatically constructed.

CLIMATOLOGY

Climatology is the study of climate, scientifically defined as weather conditions averaged over a period of time, and is a branch of the atmospheric sciences. Basic knowledge of climate can be used within shorter term weather forecasting using analog techniques such as the El Niño-Southern Oscillation (ENSO), the Madden-Julian Oscillation (MJO), the North Atlantic Oscillation (NAO), the Northern Annualar Mode (NAM), the Arctic oscillation (AO), the Northern Pacific (NP) Index, the Pacific Decadal Oscillation (PDO), and the Interdecadal Pacific Oscillation (IPO). Climate models are used for a variety of purposes from study of the dynamics of the weather and climate system to projections of future climate.

History

The earliest person to hypothesize the concept of climate change may have been the medieval Chinese scientist Shen Kuo (1031-1095 AD). Shen Kuo theorized that climates naturally shifted over an enormous span of time, after observing petrified bamboos found underground near Yanzhou (modern day Yan'an, Shaanxi province), a dry climate area unsuitable for the growth of bamboo.

Early climate researchers include Edmund Halley, who published a map of the trade winds in 1686, after a voyage to the southern hemisphere. Benjamin Franklin, in the 18th century, was the first to map the course of the Gulf Stream for use in sending mail overseas from the United States to Europe. Francis Galton invented the term *anticyclone*. Helmut Landsberg led to statistical analysis being used in climatology, which led to its evolution into a physical science.

Different Approaches

Climatology is approached in a variety of ways. Paleoclimatology seeks to reconstruct past climates by examining records such as ice cores and tree rings (dendroclimatology). Paleotempestology uses these same records to help determine hurricane frequency over millennia. The study of contemporary climates incorporates meteorological data accumulated over many years, such as records of rainfall, temperature and atmospheric composition. Knowledge of the atmosphere and its dynamics is also embodied in models, either statistical or mathematical, which help by integrating different observations and testing how they fit together. Modeling is used for understanding past, present and potential future climates. Historical climatology is the study of climate as related to human history and thus focuses only on the last few thousand years.

Climate research is made difficult by the large scale, long time periods, and complex processes which govern climate. Climate is governed by physical laws which can be expressed as differential equations. These equations are coupled and nonlinear, so that approximate solutions are obtained by using numerical methods to create global climate models. Climate is sometimes modeled as a stochastic process but this is generally accepted as an approximation to processes that are otherwise too complicated to analyze.

Indices

Scientists use climate indices based on several climate patterns (known as modes of variability) in their attempt to characterize and understand the various climate mechanisms that culminate in our daily weather. Much in the way the Dow

Jones Industrial Average, which is based on the stock prices of 30 companies, is used to represent the fluctuations in the stock market as a whole, climate indices are used to represent the essential elements of climate. Climate indices are generally devised with the twin objectives of simplicity and completeness, and each index typically represents the status and timing of the climate factor it represents. By their very nature, indices are simple, and combine many details into a generalized, overall description of the atmosphere or ocean which can be used to characterize the factors which impact the global climate system.

El Niño-Southern Oscillation

El Niño-Southern Oscillation (ENSO) is a global coupled ocean-atmosphere phenomenon. The Pacific ocean signatures, El Niño and La Niña are important temperature fluctuations in surface waters of the tropical Eastern Pacific Ocean. The name El Niño, from the Spanish for "the little boy", refers to the Christ child, because the phenomenon is usually noticed around Christmas time in the Pacific Ocean off the west coast of South America. La Niña means "the little girl". Their effect on climate in the subtropics and the tropics are profound. The atmospheric signature, the Southern Oscillation (SO) reflects the monthly or seasonal fluctuations in the air pressure difference between Tahiti and Darwin. The most recent occurrence of El Niño started in September 2006 and lasted until early 2007.

ENSO is a set of interacting parts of a single global system of coupled ocean-atmosphere climate fluctuations that come about as a consequence of oceanic and atmospheric circulation. ENSO is the most prominent known source of inter-annual variability in weather and climate around the world. The cycle occurs every two to seven years, with El Niño lasting nine months to two years within the longer term cycle, though not all areas globally are affected. ENSO has signatures in the Pacific, Atlantic and Indian Oceans. El Niño causes weather patterns which causes it to rain in specific places but not in others, this is one of many causes for the drought.

In the Pacific, during major warm events, El Niño warming extends over much of the tropical Pacific and becomes clearly linked to the SO intensity. While ENSO events are basically in phase between the Pacific and Indian Oceans, ENSO events in the Atlantic Ocean lag behind those in the Pacific by 12 to 18 months. Many of the countries most affected by ENSO events are developing countries within tropical sections of continents with economies that are largely dependent upon their agricultural and fishery sectors as a major source of food supply, employment, and foreign exchange. New capabilities to predict the onset of ENSO events in the three oceans can have global socio-economic impacts. While ENSO is a global and natural part of the Earth's climate, whether its intensity or frequency may change as a result of global warming is an important concern. Low-frequency variability has been evidenced: the quasi-decadal oscillation (QDO). Inter-decadal (ID) modulation of ENSO (from PDO or IPO) might exist. This could explain the so-called protracted ENSO of the early 1990s.

Madden-Julian Oscillation

The Madden-Julian Oscillation (MJO) is an equatorial traveling pattern of anomalous rainfall that is planetary in scale. It is characterized by an eastward progression of large regions of both enhanced and suppressed tropical rainfall, observed mainly over the Indian Ocean and Pacific Ocean. The anomalous rainfall is usually first evident over the western Indian Ocean, and remains evident as it propagates over the very warm ocean waters of the western and central tropical Pacific. This pattern of tropical rainfall then generally becomes very nondescript as it moves over the cooler ocean waters of the eastern Pacific but reappears over the tropical Atlantic and Indian Ocean. The wet phase of enhanced convection and precipitation is followed by a dry phase where convection is suppressed. Each cycle lasts approximately 30–60 days. The MJO is also known as the 30-60 day oscillation, 30-60 day wave, or intraseasonal oscillation.

North Atlantic Oscillation (NAO)

Indices of the NAO are based on the difference of normalized sea level pressure (SLP) between Ponta Delgada, Azores and Stykkisholmur/Reykjavik, Iceland. The SLP anomalies at each station were normalized by division of each seasonal mean pressure by the long-term mean (1865-1984) standard deviation. Normalization is done to avoid the series of being dominated by the greater variability of the northern of the two stations. Positive values of the index indicate stronger-than-average westerlies over the middle latitudes.

Northern Annular Mode (NAM) or Arctic Oscillation (AO)

The NAM, or AO, is defined as the first EOF of northern hemisphere winter SLP data from the tropics and subtropics. It explains 23% of the average winter (December-March) variance, and it is dominated by the NAO structure in the Atlantic. Although there are some subtle differences from the regional pattern over the Atlantic and Arctic, the main difference is larger amplitude anomalies over the North Pacific of the same sign as those over the Atlantic. This feature gives the NAM a more annular (or zonally symmetric) structure.

Northern Pacific (NP) Index

The NP Index is the area-weighted sea level pressure over the region 30N-65N, 160E-140W.

Pacific Decadal Oscillation (PDO)

The PDO is a pattern of Pacific climate variability that shifts phases on at least inter-decadal time scale, usually about 20 to 30 years. The PDO is detected as warm or cool surface waters in the Pacific Ocean, north of 20° N. During a "warm", or "positive", phase, the west Pacific becomes cool and part of the eastern ocean warms; during a "cool" or "negative" phase, the opposite pattern occurs. The mechanism by which the pattern lasts over several years has not been identified; one suggestion is that a thin layer of warm water during summer may shield deeper cold waters. A PDO signal has been reconstructed to 1661 through tree-ring chronologies in the Baja California area.

Interdecadal Pacific Oscillation (IPO)

The Interdecadal Pacific Oscillation (IPO or ID) display similar sea surface temperature (SST) and sea level pressure patterns to the PDO, with a cycle of 15–30 years, but affects both the north and south Pacific. In the tropical Pacific, maximum SST anomalies are found away from the equator. This is quite different from the quasi-decadal oscillation (QDO) with a period of 8-to-12 years and maximum SST anomalies straddling the equator, thus resembling ENSO.

Models

Climate models use quantitative methods to simulate the interactions of the atmosphere, oceans, land surface, and ice. They are used for a variety of purposes from study of the dynamics of the weather and climate system to projections of future climate. All climate models balance, or very nearly balance, incoming energy as short wave (including visible) electromagnetic radiation to the earth with outgoing energy as long wave (infrared) electromagnetic radiation from the earth. Any unbalance results in a change in the average temperature of the earth.

The most talked-about models of recent years have been those relating temperature to emissions of carbon dioxide. These models predict an upward trend in the surface temperature record, as well as a more rapid increase in temperature at higher altitudes.

Models can range from relatively simple to quite complex:

- A simple radiant heat transfer model that treats the earth as a single point and averages outgoing energy
- this can be expanded vertically (radiative-convective models), or horizontally
- finally, (coupled) atmosphere–ocean–sea ice global climate models discretise and solve the full equations for mass and energy transfer and radiant exchange.

DIFFERENCES WITH METEOROLOGY

In contrast to meteorology, which focuses on short term weather systems lasting up to a few weeks, climatology studies the frequency and trends of those systems. It studies the periodicity of weather events over years to millennia, as well as changes in long-term average weather patterns, in relation to atmospheric conditions. Climatologists, those who practice climatology, study both the nature of climates-local, regional or global-and the natural or human-induced factors that cause climates to change. Climatology considers the past and can help predict future climate change. Phenomena of climatological interest include the atmospheric boundary layer, circulation patterns, heat transfer (radiative, convective and latent), interactions between the atmosphere and the oceans and land surface (particularly vegetation, land use and topography), and the chemical and physical composition of the atmosphere.

Use in Weather Forecasting

A more complicated way of making a forecast, the analog technique requires remembering a previous weather event which is expected to be mimicked by an upcoming event. What makes it a difficult technique to use is that there is rarely a perfect analog for an event in the future. Some call this type of forecasting pattern recognition, which remains a useful method of observing rainfall over data voids such as oceans with knowledge of how satellite imagery relates to precipitation rates over land, as well as the forecasting of precipitation amounts and distribution in the future. A variation on this theme is used in Medium Range forecasting, which is known as teleconnections, when you use systems in other locations to help pin down the location of another system within the surrounding regime. One method of using teleconnections are by using climate indices such as ENSO-related phenomena.

METEOROLOGY

Meteorology is the interdisciplinary scientific study of the atmosphere that focuses on weather processes and short term forecasting (in contrast with climatology). Studies in the field stretch back millennia, though significant progress in meteorology did not occur until the eighteenth century. The nineteenth century saw breakthroughs occur after observing networks developed across several countries. Breakthroughs in weather forecasting were achieved in the latter half of the twentieth century, after the development of the computer.

Meteorological phenomena are observable weather events which illuminate and are explained by the science of meteorology. Those events are bound by the variables that exist in Earth's atmosphere: They are temperature, air pressure, water vapour, and the gradients and interactions of each variable, and how they change in time. The majority of Earth's observed weather is located in the troposphere. Different spatial scales are studied to determine how systems on local, region, and global levels impact weather and climatology. Meteorology, climatology, atmospheric physics, and atmospheric chemistry are sub-disciplines of the atmospheric sciences. Meteorology and hydrology compose the interdisciplinary field of hydrometeorology. Interactions between Earth's atmosphere and the oceans are part of coupled ocean-atmosphere studies. Meteorology has application in many diverse fields such as the military, energy production, transport, agriculture and construction.

History

In 350 BC, Aristotle wrote *Meteorology*. Aristotle is considered the founder of meteorology. One of the most impressive achievements described in the *Meteorology* is the description of what is now known as the hydrologic cycle. The Greek scientist Theophrastus compiled a book on weather forecasting, called the *Book of Signs*. The work of Theophrastus remained a dominant influence in the study of weather and in weather forecasting for nearly 2,000 years. In 25 AD, Pomponius Mela, a geographer for the Roman Empire, formalized the climatic zone system. Around the 9th century, Al-Dinawari, a Kurdish naturalist, writes the *Kitab al-Nabat* (*Book of Plants*), in which he deals with the application of meteorology to agriculture during the Muslim Agricultural Revolution. He describes the meteorological character of the sky, the planets and constellations, the sun and moon, the lunar phases indicating seasons and rain, the *anwa* (heavenly bodies of rain), and atmospheric phenomena such as winds, thunder, lightning, snow, floods, valleys, rivers, lakes, wells and other sources of water.

Research of Visual Atmospheric Phenomena

In 1021, Ibn al-Haytham (Alhazen) wrote on the atmospheric refraction of light. He showed that the twilight is due to atmospheric refraction and only begins when the Sun is 19 degrees below the horizon, and uses a complex geometric demonstration to measure the height of the Earth's atmosphere as 52,000 *passuum* (49 miles (79 km)), which is very close to the modern measurement of 50 miles (80 km). He also realized that the atmosphere also reflects light, from his observations of the sky brightening even before the Sun rises.

St. Albert the Great was the first to propose that each drop of falling rain had the form of a small sphere, and that this form meant that the rainbow was produced by light interacting with each raindrop. Roger Bacon was the first to calculate the angular size of the rainbow. He stated that the rainbow summit can not appear higher than 42 degrees above the horizon. In the late 13th century and early 14th century, Theodoric of Freiberg and Kamâl al-Dîn al-Fârisî continued the work of Ibn al-Haytham, and they were the first to give the correct explanations for the primary rainbow phenomenon. Nevertheless, Theoderic went further and also explained the secondary rainbow In 1716, Edmund Halley suggests that aurorae are caused by "magnetic effluvia" moving along the Earth's magnetic field lines.

Instruments and Classification Scales

In 1441, King Sejongs son, Prince Munjong, invented the first standardized rain gauge. These were sent throughout the Joseon Dynasty of Korea as an official tool to assess land taxes based upon a farmer's potential harvest. In 1450, Leone Battista Alberti developed a swinging-plate anemometer, and is known as the first *anemometer*. In 1607, Galileo Galilei constructs a thermoscope. In 1611, Johannes Kepler writes the first scientific treatise on snow crystals: "Strena Seu de Nive Sexangula (A New Year's Gift of Hexagonal Snow)". In 1643, Evangelista Torricelli invents the mercury barometer. In 1662, Sir Christopher Wren invented the mechanical, self-emptying, tipping bucket rain gauge. In 1714, Gabriel Fahrenheit creates a reliable scale for measuring temperature with a mercury-type thermometer. In 1742, Anders Celsius, a Swedish astronomer, proposed the 'centigrade' temperature scale, the predecessor of the current Celsius scale. In 1783, the first hair hygrometer is demonstrated by Horace-Bénédict de Saussure. In 1802-1803, Luke Howard writes *On the Modification of Clouds* in which he assigns cloud types Latin names. In 1806, Francis Beaufort introduced his system for classifying wind speeds. Near the end of the 19th century the first cloud atlases were published, including the *International Cloud Atlas*, which has remained in print ever since. The April 1960 launch of the first successful weather satellite, TIROS-1, marked the beginning of the age where weather information became available globally.

Atmospheric Composition Research

In 1648, Blaise Pascal rediscovers that atmospheric pressure decreases with height, and deduces that there is a vacuum above the atmosphere. In 1738, Daniel Bernoulli publishes *Hydrodynamics*, initiating the kinetic theory of gases and established the basic laws for the theory of gases. In 1761, Joseph Black discovers that ice absorbs heat without changing its temperature when melting. In 1772, Black's student Daniel Rutherford discovers nitrogen, which he calls *phlogisticated air*, and together they developed the phlogiston theory. In 1777, Antoine Lavoisier discovers oxygen and develops an explanation for combustion. In 1783, in Lavoisier's book *Reflexions sur le phlogistique*, he deprecates the phlogiston theory and proposes a caloric theory. In 1804, Sir John Leslie observes that a matte black surface radiates heat more effectively than a polished surface, suggesting the importance of black body radiation. In 1808, John Dalton defends caloric theory in *A New System of Chemistry* and describes how it combines with matter, especially gases; he proposes that the heat capacity of gases varies inversely with atomic weight. In 1824, Sadi Carnot analyzes the efficiency of steam engines using caloric theory; he develops the notion of a reversible process and, in postulating that no such thing exists in nature, lays the foundation for the second law of thermodynamics.

Research into Cyclones and Air Flow

In 1494, Christopher Columbus experiences a tropical cyclone, leads to the first written European account of a hurricane. In 1686, Edmund Halley presents a systematic study of the trade winds and monsoons and identifies solar heating as the cause of atmospheric motions. In 1735, an *ideal* explanation of global circulation through study of the Trade winds was written by George Hadley. In 1743, when Benjamin Franklin is prevented from seeing a lunar eclipse by a hurricane, he decides that cyclones move in a contrary manner to the winds at their periphery. Understanding the kinematics of how exactly the rotation of the Earth affects airflow was partial at first. Gaspard-Gustave Coriolis published a paper in 1835 on the energy yield of machines with rotating parts, such as waterwheels. In 1856, William Ferrel proposed the existence of a circulation cell in the mid-latitudes with air being deflected by the Coriolis force to create the prevailing westerly winds. Late in the 19th century the full extent of the large scale interaction of pressure gradient force and deflecting force that in the end causes air masses to move *along* isobars was understood. By 1912, this deflecting force was named the Coriolis effect. Just after World War II, a group of meteorologists in Norway led by Vilhelm Bjerknes developed the Norwegian cyclone model that explains the generation, intensification and ultimate decay (the life cycle) of mid-latitude cyclones, introducing the idea of fronts, that is, sharply defined boundaries between air masses. The group included Carl-Gustaf Rossby (who was the first to explain the large scale atmospheric flow in terms of fluid dynamics), Tor Bergeron (who first determined the mechanism by which rain forms) and Jacob Bjerknes.

Observation Networks and Weather Forecasting

In 1654, Ferdinando II de Medici establishes the first *weather observing* network, that consisted of meteorological stations in Florence, Cutigliano, Vallombrosa, Bologna, Parma, Milan, Innsbruck, Osnabrück, Paris and Warsaw. Collected data was centrally sent to Florence at regular time intervals. In 1832, an electromagnetic telegraph was created by Baron Schilling. The arrival of the electrical telegraph in 1837 afforded, for the first time, a practical method for quickly gathering surface weather observations from a wide area. This data could be used to produce maps of the state of the atmosphere for a region near the Earth's surface and to study how these states evolved through time. To make frequent weather forecasts based on these data required a reliable network of observations, but it was not until 1849 that the Smithsonian Institution began to establish an observation network across the United States under the leadership of Joseph Henry. Similar observation networks were established in Europe at this time. In 1854, the United Kingdom government appointed Robert FitzRoy to the new office of *Meteorological Statist to the Board of Trade* with the role of gathering weather observations at sea. FitzRoy's office became the United Kingdom Meteorological Office in 1854, the first national meteorological service in the world. The first daily weather forecasts made by FitzRoy's Office were published in *The Times* newspaper in 1860. The following year a system was introduced of hoisting storm warning cones at principal ports when a gale was expected.

Over the next 50 years many countries established national meteorological services. The India Meteorological Department (1875) was established following tropical cyclone and monsoon related famines in the previous decades. The Finnish Meteorological Central Office (1881) was formed from part of Magnetic Observatory of Helsinki University. Japan's Tokyo Meteorological Observatory, the forerunner of the Japan Meteorological Agency, began constructing surface weather maps in 1883. The United States Weather Bureau (1890) was established under the United States Department of Agriculture. The Australian Bureau of Meteorology (1906) was established by a Meteorology Act to unify existing state meteorological services.

Numerical Weather Prediction

In 1904, Norwegian scientist Vilhelm Bjerknes first argued in his paper *Weather Forecasting as a Problem in Mechanics and Physics* that it should be possible to forecast weather from calculations based upon natural laws. It was not until later in the 20th century that advances in the understanding of atmospheric physics led to the foundation of modern numerical weather prediction.

In 1922, Lewis Fry Richardson published "Weather Prediction By Numerical Process", after finding notes and derivations he worked on as an ambulance driver in World War I. He described therein how small terms in the prognostic fluid dynamics equations governing atmospheric flow could be neglected, and a finite differencing scheme in time and space could be devised, to allow numerical prediction solutions to be found. Richardson envisioned a large auditorium of thousands of people performing the calculations and passing them to others. However, the sheer number of calculations required was too large to be completed without the use of computers, and the size of the grid and time steps led to unrealistic results in deepening systems. It was later found, through numerical analysis, that this was due to numerical instability.

Starting in the 1950s, numerical forecasts with computers became feasible. The first weather forecasts derived this way used barotropic (that means, single-vertical-level) models, and could successfully predict the large-scale movement of midlatitude Rossby waves, that is, the pattern of atmospheric lows and highs.

In the 1960s, the chaotic nature of the atmosphere was first observed and mathematically described by Edward Lorenz, founding the field of chaos theory. These advances have led to the current use of ensemble forecasting in most major forecasting centers, to take into account uncertainty arising from the chaotic nature of the atmosphere. In recent years, climate models have been developed that feature a resolution comparable to older weather prediction models. These climate models are used to investigate long-term climate shifts, such as what effects might be caused by human emission of greenhouse gases.

Meteorologists

Meteorologists are scientists who study meteorology. Meteorologists work in government agencies, private consulting and research services, industrial enterprises, utilities, radio and television stations, and in education. In the United States, meteorologists held about 9,400 jobs in 2009. Meteorologists are best-known for forecasting the weather. Many radio and television weather forecasters are professional meteorologists, while others are merely reporters with no formal meteorological training. The American Meteorological Society and National Weather Association issue "Seals of Approval" to weather broadcasters who meet certain requirements.

Equipment

Each science has its own unique sets of laboratory equipment. In the atmosphere, there are many things or qualities of the atmosphere that can be measured. Rain, which can be observed, or seen anywhere and anytime was one of the first ones to be measured historically. Also, two other accurately measured *qualities* are wind and humidity. Neither of these can be *seen* but can be felt. The devices to measure these three sprang up in the mid-15th century and were respectively the rain gauge, the anemometer, and the hygrometer.

Sets of surface measurements are important data to meteorologists. They give a snapshot of a variety of weather conditions at one single location and are usually at a weather station, a ship or a weather buoy. The measurements taken at a weather station can include any number of atmospheric observables. Usually, temperature, pressure, wind measurements, and humidity are the variables that are measured by a thermometer, barometer, anemometer, and hygrometer, respectively. Upper air data are of crucial importance for weather forecasting. The most widely used technique is launches of radiosondes. Supplementing the radiosondes a network of aircraft collection is organized by the World Meteorological Organization. Remote sensing, as used in meteorology, is the concept of collecting data from remote weather events and subsequently producing weather information. The common types of remote sensing are Radar, Lidar, and satellites (or photogrammetry). Each collects data about the atmosphere from a remote location and, usually, stores the data where the instrument is located. RADAR and LIDAR are not passive because both use EM radiation to illuminate a specific portion of the atmosphere. Weather satellites along with more general-purpose Earth-observing satellites circling the earth at various altitudes have become an indispensable tool for studying a wide range of phenomena from forest fires to El Niño.

Spatial Scales

In the study of the atmosphere, meteorology can be divided into distinct areas of emphasis depending on the temporal scope and spatial scope of interest. At one extreme of this scale is climatology. In the timescales of hours to days, meteorology separates into micro-, meso-, and synoptic scale meteorology. Respectively, the geospatial size of each of these three scales relates directly with the appropriate timescale. Other subclassifications are available based on the need by or by the unique, local or broad effects that are studied within that sub-class.

Microscale

Microscale meteorology is the study of atmospheric phenomena of about 1 km or less. Individual thunderstorms, clouds, and local turbulence caused by buildings and other obstacles, such as individual hills fall within this category.

Mesoscale

Mesoscale meteorology is the study of atmospheric phenomena that has horizontal scales ranging from microscale limits to synoptic scale limits and a vertical scale that starts at the Earth's surface and includes the atmospheric boundary layer, troposphere, tropopause, and the lower section of the stratosphere. Mesoscale timescales last from less than a day to the lifetime of the event, which in some cases can be weeks. The events typically of interest are thunderstorms, squall lines, fronts, precipitation bands in tropical and extratropical cyclones, and topographically generated weather systems such as mountain waves and sea and land breezes.

Synoptic Scale

Synoptic scale meteorology is generally large area dynamics referred to in horizontal coordinates and with respect to time. The phenomena typically described by synoptic meteorology include events like extratropical cyclones, baroclinic troughs and ridges, frontal zones, and to some extent jet streams. All of these are typically given on weather maps for a specific time. The minimum horizontal scale of synoptic phenomena are limited to the spacing between surface observation stations.

Global Scale

Global scale meteorology is study of weather patterns related to the transport of heat from the tropics to the poles. Also, very large scale oscillations are of importance. Those oscillations have time periods typically longer than a full annual seasonal cycle, such as ENSO, PDO, MJO, etc. Global scale pushes the thresholds of the perception of meteorology into climatology. The traditional definition of climate is pushed in to larger timescales with the further understanding of how the global oscillations cause both climate and weather disturbances in the synoptic and mesoscale timescales. Numerical Weather Prediction is a main focus in understanding air-sea interaction, tropical meteorology, atmospheric predictability, and tropospheric/stratospheric processes. Currently (2007) Naval Research Laboratory in Monterey produces the atmospheric model called NOGAPS, a global scale atmospheric model, this model is run operationally at Fleet Numerical Meteorology and Oceanography Centre. There are several other global atmospheric models.

SOME METEOROLOGICAL PRINCIPLES

Boundary Layer Meteorology

Boundary layer meteorology is the study of processes in the air layer directly above Earth's surface, known as the atmospheric boundary layer (ABL). The effects of the surface – heating, cooling, and friction – cause turbulent mixing within the air layer. Significant fluxes of heat, matter, or momentum on time scales of less than a day are advected by turbulent motions. Boundary layer meteorology includes the study of all types of surface-atmosphere boundary, including ocean, lake, urban land and non-urban land.

Dynamic Meteorology

Dynamic meteorology generally focuses on the fluid dynamics of the atmosphere. The idea of air parcel is used to define the smallest element of the atmosphere, while ignoring the discrete molecular and chemical nature of the atmosphere. An air parcel is defined as a point in the fluid continuum of the atmosphere. The fundamental laws of fluid dynamics, thermodynamics, and motion are used to study the atmosphere. The physical quantities that characterize the state of the atmosphere are temperature, density, pressure, etc. These variables have unique values in the continuum.

APPLICATIONS

Weather Forecasting

Weather forecasting is the application of science and technology to predict the state of the atmosphere for a future time and a given location. Human beings have attempted to predict the weather informally for millennia, and formally since at least the nineteenth century. Weather forecasts are made by collecting quantitative data about the current state of the atmosphere and using scientific understanding of atmospheric processes to project how the atmosphere will evolve. Once an all human endeavor based mainly upon changes in barometric pressure, current weather conditions, and sky condition, forecast models are now used to determine future conditions. Human input is still required to pick the best possible forecast model to base the forecast upon, which involves pattern recognition skills, teleconnections, knowledge of model performance, and knowledge of model biases. The chaotic nature of the atmosphere, the massive computational power required to solve the equations that describe the atmosphere, error involved in measuring the initial conditions, and an incomplete understanding of atmospheric processes mean that forecasts become less accurate as the difference in current time and the time for which the forecast is being made (the *range* of the forecast) increases. The use of ensembles and model consensus help narrow the error and pick the most likely outcome.

There are a variety of end uses to weather forecasts. Weather warnings are important forecasts because they are used to protect life and property. Forecasts based on temperature and precipitation are important to agriculture, and therefore to commodity traders within stock markets. Temperature forecasts are used by utility companies to estimate demand over coming days. On an everyday basis, people use weather forecasts to determine what to wear on a given day. Since outdoor activities are severely curtailed by heavy rain, snow and the wind chill, forecasts can be used to plan activities around these events, and to plan ahead and survive them.

Aviation Meteorology

Aviation meteorology deals with the impact of weather on air traffic management. It is important for air crews to understand the implications of weather on their flight plan as well as their aircraft, as noted by the Aeronautical Information Manual :

The effects of ice on aircraft are cumulative-thrust is reduced, drag increases, lift lessens, and weight increases. The results are an increase in stall speed and a deterioration of aircraft performance. In extreme cases, 2 to 3 inches of ice can form on the leading edge of the airfoil in less than 5 minutes. It takes but 1/2 inch of ice to reduce the lifting power of some aircraft by 50 percent and increases the frictional drag by an equal percentage.

Agricultural Meteorology

Meteorologists, soil scientists, agricultural hydrologists, and agronomists are persons concerned with studying the effects of weather and climate on plant distribution, crop yield, water-use efficiency, phenology of plant and animal development, and the energy balance of managed and natural ecosystems. Conversely, they are interested in the role of vegetation on climate and weather.

Hydrometeorology

Hydrometeorology is the branch of meteorology that deals with the hydrologic cycle, the water budget, and the rainfall statistics of storms. A hydrometeorologist prepares and issues forecasts of accumulating (quantitative) precipitation, heavy rain, heavy snow, and highlights areas with the potential for flash flooding. Typically the range of knowledge that is required overlaps with climatology, mesoscale and synoptic meteorology, and other geosciences.

Nuclear Meteorology

Nuclear meteorology investigates the distribution of radioactive aerosols and gases in the atmosphere.

Maritime Meteorology

Maritime meteorology deals with air and wave forecasts for ships operating at sea. Organizations such as the Ocean Prediction Centre, Honolulu National Weather Service forecast office, United Kingdom Met Office, and JMA prepare high seas forecasts for the world's oceans.

PALEOCLIMATOLOGY

Paleoclimatology (also Palaeoclimatology) is the study of climate change taken on the scale of the entire history of Earth. It uses a variety of proxy records from ice sheets, tree rings, sediment, corals, shells and rocks to determine the past state of the climate system on Earth. Paleoclimatology has wider implications for climate change today. Scientists often consider past changes in environment and biodiversity to reflect on the current situation, and specifically the impact of climate on mass extinctions and biotic recovery.

Reconstructing Ancient Climates

Paleoclimatologists employ a wide variety of techniques to deduce ancient climates.

Ice

Mountain Glaciers and the polar ice caps/ice sheets are a widely employed source of data in paleoclimatology. Recent ice coring projects in the ice caps of Greenland and Antarctica have yielded data going back several hundred thousand years—over 800,000 years in the case of the EPICA project.

- Air trapped within fallen snow becomes encased in tiny bubbles as the snow is compressed into ice in the glacier under the weight of later years' snow. This trapped air has proven a tremendously valuable source for direct measurement of the composition of air from the time the ice was formed.
- Layering can be observed due to seasonal pauses in ice accumulation and can be used to establish chronology; associating specific depths of the core with ranges of time.
- Changes in the layering thickness can be used to determine changes in precipitation or temperature.
- Oxygen-18 quantity changes (ä^{18}O) in ice layers represent changes in average ocean surface temperature. Water molecules containing the heavier O-18 evaporate at a higher temperature than water molecules containing the normal Oxygen-16 isotope. The ratio of O-18 to O-16 will be higher as temperature increases and less as temperature decreases. Various cycles in those isotope ratios have been detected.
- Pollen has been observed in the ice cores and can be used to understand which plants were present as the layer formed. Pollen is produced in abundance and its distribution is typically well understood. A pollen count for a specific layer can be produced by observing the total amount of pollen categorized by type (shape) in a controlled sample of that layer. Changes in plant frequency over time can be plotted through statistical analysis of pollen counts in the core. Knowing which

plants were present leads to an understanding of precipitation and temperature, and types of fauna present. Palynology includes the study of pollen for these purposes.

- Volcanic ash is contained in some layers, and can be used to establish the time of the layer's formation. Each volcanic event distributed ash with a unique set of properties (shape and colour of particles, chemical signature). Establishing the ash's source will establish a range of time to associate with layer of ice.

Dendroclimatology

Climatic information can be obtained through an understanding of changes in tree growth. Generally, trees respond to changes in climatic variables by speeding up or slowing down growth, which in turn is generally reflected a greater or lesser thickness in growth rings. Different species, however, respond to changes in climatic variables in different ways. A tree-ring record is established by compiling information from many living trees in a specific area. Older intact wood that has escaped decay can extend the time covered by the record by matching the ring depth changes to contemporary specimens. Using this method some areas have tree-ring records dating back a few thousand years. Older wood not connected to a contemporary record can be dated generally with radiocarbon techniques. A tree-ring record can be used to produce information regarding precipitation, temperature, hydrology, and fire corresponding to a particular area. On a longer time scale, geologists must refer to the sedimentary record for data.

Sedimentary Content

- Sediments, sometimes lithified to form rock, may contain remnants of preserved vegetation, animals, plankton or pollen, which may be characteristic of certain climatic zones.
- Biomarker molecules such as the alkenones may yield information about their temperature of formation.
- Chemical signatures, particularly Mg/Ca ratio of calcite in Foraminifera tests, can be used to reconstruct past temperature.

- Isotopic ratios can provide further information. Specifically, the $\delta^{18}O$ record responds to changes in temperature and ice volume, and the $\delta^{13}C$ record reflects a range of factors, which are often difficult to disentangle.

Sedimentary Facies

On a longer time scale, the rock record may show signs of sea level rise and fall; further, features such as "fossilised" sand dunes can be identified. Scientists can get a grasp of long term climate by studying sedimentary rock going back billions of years. The division of earth history into separate periods is largely based on visible changes in sedimentary rock layers that demarcate major changes in conditions. Often these include major shifts in climate.

Corals

Coral "rings" are similar to tree rings, except they respond to different things, such as the water temperature and wave action. From this source, certain equipment can be used to derive the sea surface temperature and water salinity from the past few centuries. The $ä^{18}O$ of coraline red algae provides a useful proxy of sea surface temperature at high latitudes, where many traditional techniques are limited.

Limitations

All records decrease in utility back in time. The oldest ice core taken was from the Antarctic and dates to 800,000 years old. An international effort is currently being made in the same location to core to 1.2 million years ago. The deep marine record, the source of most isotopic data, only exists on oceanic plates, which are eventually subducted — the oldest remaining material is 200 million years old. Older sediments are also more prone to corruption by diagenesis. Resolution and confidence in the data decrease over time.

Planet's Timeline

Knowledge of precise climatic events decreases as the record goes further back in time. Some notable events are noted below, with a timescale for context.

- Faint young Sun paradox (start)
- Snowball Earth (~2400Mya)

- Andean-Saharan glaciation (~450Mya)
- Permian-Triassic extinction event (251.4Mya)
- Paleocene-Eocene Thermal Maximum (Paleocene-Eocene, 55Mya)
- Younger Dryas/The Big Freeze (~11Kya)
- Holocene climatic optimum (~7-3Kya)
- Climate changes of 535-536 (535-536 AD)
- Medieval warm period (900-1300)
- Little ice age (1300-1800)
- Year Without a Summer (1816).

HISTORY OF THE ATMOSPHERE

Earliest Atmosphere

The outgassings of the Earth was stripped away by solar winds early in the history of the planet until a steady state was established, the first atmosphere. Based on today's volcanic evidence, this atmosphere would have contained 80% water vapour, 10% carbon dioxide, 5 to 7% hydrogen sulfide, and smaller amounts of nitrogen, carbon monoxide, hydrogen, methane and inert gases. A major rainfall led to the buildup of a vast ocean, enriching the other agents, first carbon dioxide and later nitrogen and inert gases. A major part of carbon dioxide exhalations were soon dissolved in water and built up carbonate sediments.

Second Atmosphere

As early as 3.8 billion years ago, water related sediments have been found. About 3.4 billion years ago, nitrogen was the major part of the then stable second atmosphere. An influence of life has to be taken into account rather soon since hints on early life forms are to be found as early as 3.5 billion years ago. The fact that this is not in line with the — compared to today 30% lower — solar radiance of the early sun has been described as the faint young Sun paradox. The geological record, however, shows a continually relatively warm surface during the complete early temperature record of the earth with the exception of one cold glacial phase about 2.4 billion years ago. In the late Archean Era an oxygen containing atmosphere began to develop from photosynthesizing algae. The early basic carbon isotopy is very much in line with what is found today As Jan Veizer assumed that not only did we have life as far back as we had rocks, but there was as much life then as today and the fundamental features of the carbon cycle were established as early as 4 billion years ago.

Third Atmosphere

The accretion of continents about 3.5 billion years ago added plate tectonics, constantly rearranging the continents and also shaping long-term climate evolution by allowing the transfer of carbon dioxide to large land-based carbonate storages. Free oxygen did not exist until about 1.7 billion years ago and this can be seen with the development of the red beds and the end of the banded iron formations. This signifies a shift from a reducing atmosphere to an oxidising atmosphere. O_2 showed major ups and downs until reaching a steady state of more than 15%. The following time span was the Phanerozoic, during which oxygen-breathing metazoan life forms began to appear.

CLIMATE DURING GEOLOGICAL AGES

Precambrian Climate

In the first three quarters of the Earth's history, only one major glaciation is to be found in the geological record. Since about 950 million years ago, the Earth's climate has varied regularly between large-scale or just polar cap wide glaciation and extensively tropical climates. The time scale for this variation is roughly 140 million years and may be related to Earth's motion into and out of galactic spiral arms and compared to the previous time, significantly reduced solar wind.

The climate of the late Precambrian showed some major glaciation events spreading over much of the earth. At this time the continents were bunched up in the Rodinia supercontinent.

Massive deposits of tillites are found and anomalous isotopic signatures are found, which gave rise to the Snowball Earth hypothesis. As the Proterozoic Eon drew to a close, the Earth started to warm up. By the dawn of the Cambrian and the Phanerozoic, life forms were abundant in the Cambrian explosion with average global temperatures of about 22 °C.

Phanerozoic Climate

Major drivers for the preindustrial ages have been variations of the sun, volcanic ashes and exhalations, relative movements of the earth towards the sun and tectonically induced effects as for major sea currents, watersheds and ocean oscillations. In the early Phanerozoic, increased atmospheric carbon dioxide concentrations have been linked to driving or amplifying increased global temperatures. Royer et al. 2004 found a climate sensitivity for the rest of the Phanerozoic which was calculated to be similar to today's modern range of values.

The difference in global mean temperatures between a fully glacial Earth and an ice free Earth is estimated at approximately 10 °C, though far larger changes would be observed at high latitudes and smaller ones at low latitudes. One requirement for the development of large scale ice sheets seems to be the arrangement of continental land masses at or near the poles.

The constant rearrangement of continents by plate tectonics can also shape long-term climate evolution. However, the presence or absence of land masses at the poles is not sufficient to guarantee glaciations or exclude polar ice caps. Evidence exists of past warm periods in Earth's climate when polar land masses similar to Antarctica were home to deciduous forests rather than ice sheets.

The relatively warm local minimum between Jurassic and Cretaceous goes along with widespread tectonic activity, e.g. the breakup of supercontinents. Superimposed on the long-term evolution between hot and cold climates have been many short-term fluctuations in climate similar to, and sometimes more severe than, the varying glacial and interglacial states of the present ice age. Some of the most severe fluctuations, such as the Paleocene-Eocene Thermal Maximum, may be related to rapid climate changes due to sudden collapses of natural methane clathrate reservoirs in the oceans.

A similar, single event of induced severe climate change after a meteorite impact has been proposed as reason for the Cretaceous-Tertiary extinction event. Other major thresholds are the Permian-Triassic, and Ordovician-Silurian extinction events with various reasons suggested.

Quaternary Sub-era

The Quaternary sub-era includes the current climate. There has been a cycle of ice ages for the past 2.2–2.1 million years (starting before the Quaternary in the late Neogene Period). Note in the graphic on the right the strong 120,000-year periodicity of the cycles, and the striking asymmetry of the curves. This asymmetry is believed to result from complex interactions of feedback mechanisms. It has been observed that ice ages deepen by progressive steps, but the recovery to interglacial conditions occurs in one big step.

CONTROLLING FACTORS

Short Term (10^4 to 10^6 years)

Geologically short-term (<120,000 year) temperatures are believed to be driven by orbital factors amplified by changes in greenhouse gases. The arrangements of land masses on the Earth's surface are believed to influence the effectiveness of these orbital forcing effects.

Medium Term (10^6 to 10^8 years)

Continental drift affects the thermohaline circulation, which transfers heat between the equatorial regions and the poles, as does the extent of polar ice coverage. The timing of ice ages throughout geologic history is in part controlled by the position of the continental plates on the surface of the Earth. When landmasses are concentrated near the polar regions, there is an increased chance for snow and ice to accumulate. Small changes in solar energy can tip the balance between summers in which the winter snow mass completely melts and summers in which the winter snow persists until the following winter.

Comparisons of plate tectonic continent reconstructions and paleoclimatic studies show that the Milankovitch cycles have the greatest effect during geologic eras when landmasses have been concentrated in polar regions, as is the case today. Today, Greenland, Antarctica, and the northern portions of Europe, Asia, and North America are situated such that a minor change in solar energy will tip the balance between year-round snow/ice preservation and complete summer melting. The presence of snow and ice is a well-understood positive feedback mechanism for climate. The Earth today is considered to be prone to ice age glaciations. Another proposed factor in long term temperature change is the Uplift-Weathering Hypothesis, first put forward by T. C. Chamberlin in 1899 and later independently proposed in 1988 by Maureen Raymo and colleagues, where upthrusting mountain ranges expose minerals to weathering resulting in their chemical conversion to carbonates thereby removing CO_2 from the atmosphere and cooling the earth. Others have proposed similar effects due to changes in average water table levels and consequent changes in sub-surface biological activity and PH levels.

Long Term (10^8 to 10^9 years)

It has been proposed that long term galactic motions of the sun have a major influence on the Earth's climate. There are two principal motions, the first and most significant is the orbit of the sun around the galactic centre with a period of the order of 240 million years. Since this period is different from the rotation period of the galactic spiral arms, the sun, and the earth with it, will periodically pass through the arms (estimates of the period are uncertain and vary from 143 million years to 176 million years). The second is an oscillatory bobbing motion, similar to a floating buoy, which will periodically take the sun through the galactic disc. The period of this bobbing motion is 67 million years, so a pass through the galactic plane will occur every 33 million years. The causal link between these galactic motions and climate is unclear but one (controversial) postulate is the effect that entering a denser region of the galaxy will have on increasing the cosmic ray flux (CRF). This theory has been criticised, both for overstating the correlation with CRF and for failing to propose a believable mechanism that would allow CRF to drive temperature. The claims by Henrik Svensmark that CRF also strongly affects short term climate changes is even more controversial and has been challenged by many.

It has also been suggested that there is some correlation between these galactic cycles and geological periods. The reason for this is postulated to be that the earth experiences many more impact events while passing through high density regions of the galaxy. Both the climate changes and sudden impacts may cause, or contribute to, extinction events.

Very Long Term (10^9 years or more)

Jan Veizer and Nir Shaviv have proposed the interaction of cosmic rays, solar wind and the various magnetic fields to explain the long term evolution of earths climate. According to Shaviv, the early sun had emitted a stronger solar wind with a protective effect against cosmic rays. In that early age, a moderate greenhouse effect comparable to today's would have been sufficient to explain an ice free earth and the faint young sun paradox. The solar minimum around 2.4 billion years ago is consistent with an established cosmic ray flux modulation by a variable star formation rate in the Milky Way and there is also a hint of an extinction event at this time. Within the last billion years the solar wind has significantly diminished. It is only within this more recent time that passages of the heliosphere through the spiral arms of the galaxy have been able to gain a strong and regularly modulating influence as described above.

Over the very long term the energy output of the sun has gradually increased, on the order of 5% per billion (10^9) years, and will continue to do so until it reaches the end of its current phase of stellar evolution.

4

The Biosphere

Our biosphere is the global sum of all ecosystems. It can also be called the zone of life on Earth, a closed (apart from solar and cosmic radiation) and self-regulating system. From the broadest biophysiological point of view, the biosphere is the global ecological system integrating all living beings and their relationships, including their interaction with the elements of the lithosphere, hydrosphere and atmosphere.

The biosphere is postulated to have evolved, beginning through a process of biogenesis or biopoesis, at least some 3.5 billion years ago. In a broader sense; biospheres are any closed, self-regulating systems containing ecosystems; including artifical ones such as Biosphere 2 and BIOS-3; and, potentially, ones on other planets or moons.

ORIGIN AND USE OF THE TERM

The term "biosphere" was coined by geologist Eduard Suess in 1875, which he defined as: "The place on Earth's surface where life dwells." While this concept has a geological origin, it is an indication of the impact of both Darwin and Maury on the earth sciences.

The biosphere's ecological context comes from the 1920s, preceding the 1935 introduction of the term "ecosystem" by Sir Arthur Tansley. Vernadsky defined ecology as the science of the biosphere. It is an interdisciplinary concept for integrating astronomy, geophysics, meteorology, biogeography, evolution, geology, geochemistry, hydrology and, generally speaking, all life and earth sciences.

Gaia Hypothesis

The concept that the biosphere is itself a living organism, either actually or metaphorically, is known as the Gaia hypothesis. James Lovelock, an atmospheric scientist from the United Kingdom, proposed the Gaia hypothesis to explain how biotic and abiotic factors interact in the biosphere. This hypothesis considers Earth itself a kind of living organism. Its atmosphere, geosphere, and hydrosphere are cooperating systems that yield a biosphere full of life. In the early 1970s, Lynn Margulis, a microbiologist from the United States, added to the hypothesis, specifically noting the ties between the biosphere and other Earth systems. For example, when carbon dioxide levels increase in the atmosphere, plants grow more quickly. As their growth continues, they remove more and more carbon dioxide from the atmosphere.

There are many scientists are now involved in new fields of study that examine interactions between biotic and abiotic factors in the biosphere, such as geobiology and geomicrobiology. Ecosystems occur when communities and their physical environment work together as a system. The difference between this and a biosphere is simple, the biosphere is everything in general terms.

EXTENT OF EARTH'S BIOSPHERE

Every part of the planet, from the polar ice caps to the Equator, supports life of some kind. Recent advances in microbiology have demonstrated that microbes live deep beneath the Earth's terrestrial surface, and that the total mass of microbial life in so-called "uninhabitable zones" may, in biomass, exceed all animal and plant life on the surface. The actual thickness of the biosphere on earth is difficult to measure. Birds typically fly at altitudes of 650 to 1,800 meters, and fish that live deep underwater can be found down to-8,372 meters in the Puerto Rico Trench. There are more extreme examples for life on the planet: Rüppell's Vulture has been found at altitudes of 11,300 meters; Bar-headed Geese migrate at altitudes of at least 8,300 meters (over Mount Everest); Yaks live at elevations between 3,200 to 5,400 meters above sea level; mountain goats live up to 3,050 meters. Herbivorous animals at these elevations depend on lichens, grasses, and herbs. Microscopic organisms live at such extremes that, taking them into consideration puts the thickness of the biosphere much greater.

Culturable microbes have been found in the Earth's upper atmosphere as high as 41 km (25 mi). It is unlikely, however, that microbes are active at such altitudes, where temperatures and air pressure are extremely low and ultraviolet radiation very high. More likely these microbes were brought into the upper atmosphere by winds or possibly volcanic eruptions. Barophilic marine microbes have been found at more than 10 km (6 mi) depth in the Marianas Trench. Microbes are not limited to the air, water or the Earth's surface. Culturable thermophilic microbes have been extracted from cores drilled more than 5 km (3 mi) into the Earth's crust in Sweden, from rocks between 65-75 °C. Temperature increases with increasing depth into the Earth's crust. The speed at which the temperature increases depends on many factors, including type of crust (continental vs. oceanic), rock type, geographic location, etc. The upper known limit of microbial is 122 °C (*Methanopyrus kandleri* Strain 116), and it is likely that the limit of life in the "deep biosphere" is defined by temperature rather than absolute depth.

Our biosphere is divided into a number of biomes, inhabited by broadly similar flora and fauna. On land, biomes are separated primarily by latitude. Terrestrial biomes lying within the Arctic and Antarctic Circles are relatively barren of plant and animal life, while most of the more populous biomes lie near the equator. Terrestrial organisms in temperate and Arctic biomes have relatively small amounts of total biomass, smaller energy budgets, and display prominent adaptations to cold, including world-spanning migrations, social adaptations, homeothermy, estivation and multiple layers of insulation.

BIOGEOGRAPHY

Biogeography is the study of the distribution of species (biology) spatially (geography) and temporally (history). Biogeography aims to reveal where organisms live, at what abundance, and why (or why not) they are found in a geographical area. The patterns of species distribution across geographical areas can usually be explained through a combination of historical factors such as: speciation; extinction; continental drift; glaciation, and associated variations in sea level, river routes, habitat; and river capture; in combination with the geographic constraints of landmass areas and isolation; and the available ecosystem energy supplies.

Over periods of ecological changes, biogeography includes the study of plant and animal species in: their past and/or present living *refugium* habitat; their interim living sites; and/ or their survival locales. As writer David Quammen put it, "...biogeography does more than ask *Which species?* and *Where.* It also asks *Why?* and, what is sometimes more crucial, *Why not?.*"

Modern biogeography often employs the use of Geographic Information Systems (GIS), to understand the factors affecting organism distribution, and to predict future trends in organism distribution. Often mathematical models and GIS are employed to solve ecological problems that have a spatial aspect to them.

History

The scientific theory of biogeography grows out of the work of Hewett Cottrell Watson (1804–1881), Alfred Russel Wallace (1823–1913) and other early evolutionary scientists. Wallace studied the distribution of flora and fauna in the Malay Archipelago in the 19th century. With the exception of Wallace and a few others, prior to the publication of *The Theory of Island Biogeography* by Robert MacArthur and E.O. Wilson in 1967 the field of biogeography was seen as a primarily historical one and as such the field was seen as a purely descriptive one. MacArthur and Wilson changed this perception, and showed that the species richness of an area could be predicted in terms of such factors as habitat area, immigration rate and extinction rate. This gave rise to an interest in island biogeography. The application of island biogeography theory to habitat fragments spurred the development of the fields of conservation biology and landscape ecology.

Classic biogeography has been expanded by the development of molecular systematics, creating a new discipline known as phylogeography. This development allowed scientists to test theories about the origin and dispersal of populations, such as island endemics. For example, while classic biogeographers were able to speculate about the origins of species in the Hawaiian Islands, phylogeography allows them to test theories of relatedness between these populations and putative source populations in Asia and North America.

Paleobiogeography

Paleobiogeography goes one step further to include paleogeographic data and considerations of plate tectonics. Using molecular analyses and corroborated by fossils, it has been possible to demonstrate that perching birds evolved first in the region of Australia or the adjacent Antarctic (which at that time lay somewhat further north and had a temperate climate). From there, they spread to the other Gondwanan continents and Southeast Asia-the part of Laurasia then closest to their origin of dispersal – in the late Paleogene, before achieving a global distribution in the early Neogene. Not knowing the fact that at the time of dispersal, the Indian Ocean was much narrower than it is today, and that South America was closer to the Antarctic, one would be hard pressed to explain the presence of many "ancient" lineages of perching birds in Africa, as well as the mainly South American distribution of the suboscines.

Classification

Biogeography is a synthetic science, related to geography, biology, soil science, geology, climatology, ecology and evolution. Some fundamental concepts in biogeography include:

- evolution (change in genetic composition of a population)
- extinction (disappearance of a species)
- dispersal (movement of populations away from their point of origin, related to migration)
- geodispersal the erosion of barriers to biotic dispersal and gene flow, that permit range expansion and the merging of previously isolated biotas.
- range and distribution
- endemic areas
- vicariance the formation of barriers to biotic dispersal and gene flow, that tend to subdivide species and biotas, leading to speciation and extinction.

PALEONTOLOGY

Paleontology is the study of prehistoric life, including organisms' evolution and interactions with each other and their environments (their paleoecology). As a "historical science" it tries to explain causes rather than conduct experiments to observe effects. Paleontological observations have been documented as far back as the 5th century BC. The science became established in the 18th century as a result of Georges Cuvier's work on comparative anatomy, and developed rapidly in the 19th century. Fossils found in China since the 1990s have provided new information about the earliest evolution of animals, early fish, dinosaurs and the evolution of birds and mammals. Paleontology lies on the border between biology and geology, and shares with archaeology a border that is difficult to define. It now uses techniques drawn from a wide range of sciences, including biochemistry, mathematics and engineering. As knowledge has increased, paleontology has developed specialized subdivisions, some of which focus on different types of fossil organisms while others study ecological and environmental history, such as ancient climates.

Body fossils and trace fossils are the principal types of evidence about ancient life, and geochemical evidence has helped to decipher the evolution of life before there were organisms large enough to leave fossils. Estimating the dates of these remains is essential but difficult: sometimes adjacent rock layers allow radiometric dating, which provides absolute dates that are accurate to within 0.5%, but more often paleontologists have to rely on relative dating by solving the "jigsaw puzzles" of biostratigraphy.

Classifying ancient organisms is also difficult, as many do not fit well into the Linnean taxonomy that is commonly used for classifying living organisms, and paleontologists more often use cladistics to draw up evolutionary "family trees". The final quarter of the 20th century saw the development of molecular phylogenetics, which investigates how closely organisms are related by measuring how similar the DNA is in their genomes. Molecular phylogenetics has also been used to estimate the dates when species diverged, but there is controversy about the reliability of the molecular clock on which such estimates depend.

Use of all these techniques has enabled paleontologists to discover much of the evolutionary history of life, almost all the way back to when Earth became capable of supporting life, about 3,800 million years ago. For about half of that time the only life was single-celled micro-organisms, mostly in microbial mats that formed ecosystems only a few millimeters thick. Earth's atmosphere originally contained virtually no oxygen, and its oxygenation began about 2,400 million years ago. This may have caused an accelerating increase in the diversity and complexity of life, and early multicellular plants and fungi have been found in rocks dated from 1,700 to 1,200 million years ago. The earliest multicellular animal fossils are much later, from about 580 million years ago, but animals diversified very rapidly and there is a lively debate about whether most of this happened in a relatively short Cambrian explosion or started earlier but has been hidden by lack of fossils. All of these organisms lived in water, but plants and invertebrates started colonizing land from about 490 million years ago and vertebrates followed them about 370 million years ago. The first dinosaurs appeared about 230 million years ago and birds evolved from one dinosaur group about 150 million years ago. During the time of the dinosaurs, mammals' ancestors survived only as small, mainly nocturnal insectivores, but after the non-avian dinosaurs became extinct in the Cretaceous–Tertiary extinction event 65 million years ago mammals diversified rapidly. Flowering plants appeared and rapidly diversified between 130 million years ago and 90 million years ago, possibly helped by coevolution with pollinating insects. Social insects appeared around the same time and, although they have relatively few species, now form over 50% of the total mass of all insects. The upright-walking common ancestor of humans and chimpanzees *Sahelanthropus tchadensis* appeared around 6 to 7 million years ago, and anatomically modern humans appeared under 200,000 years ago. The course of evolution has been changed several times by mass extinctions that wiped out previously dominant groups and allowed other to rise from obscurity to become major components of ecosystems.

Definition

The simplest definition is "the study of ancient life". Paleontology seeks information about several aspects of past organisms: "their identity and origin, their environment and evolution, and what they can tell us about the Earth's organic and inorganic past".

A Historical Science

Paleontology is one of the historical sciences, along with archaeology, geology, biology, astronomy, cosmology, philology and history itself. This means that it aims to describe phenomena of the past and reconstruct their causes. Hence it has three main elements: description of the phenomena; developing a general theory about the causes of various types of change; and applying those theories to specific facts.

When trying to explain past phenomena, paleontologists and other historical scientists often construct a set of hypotheses about the causes and then look for a "smoking gun", a piece of evidence which indicates that one of the hypotheses is a better explanation than the others. Sometimes the "smoking gun" is discovered by a fortunate accident during other research, for example the discovery by Luis Alvarez and Walter Alvarez of an iridium-rich layer at the Cretaceous–Tertiary boundary made asteroid impact and volcanism the most favored explanations for the Cretaceous–Tertiary extinction event.

The other main type of science is experimental science, which is often said to work by conducting experiments to *disprove* hypotheses about the workings and causes of natural phenomena – note that this approach cannot prove a hypothesis is correct, since some later experiment may disprove it. However, when confronted with totally unexpected phenomena, such as the first evidence for invisible radiation, experimental scientists often use the same approach as historical scientists: construct a set of hypotheses about the causes and then look for a "smoking gun".

Related Sciences

Paleontology lies on the boundary between biology and geology since paleontology focuses on the record of past life but its main source of evidence is fossils, which are found in rocks. For historical reasons paleontology is part of the geology departments of many universities, because in the 19th century and early 20th century geology departments found paleontological evidence important for estimating the ages of rocks while biology departments showed little interest. Paleontology also has some overlap with archaeology, which primarily works with objects made by humans and with human remains, while paleontologists are interested in the characteristics and evolution of humans as organisms. When dealing with evidence about humans, archaeologists and paleontologists may work together – for example paleontologists might identify animal or plant fossils around an archaeological site, to discover what the people who lived there ate; or they might analyze the climate at the time when the site was inhabited by humans.

In addition paleontology often uses techniques derived from other sciences, including biology, ecology, chemistry, physics and mathematics. For example geochemical signatures from rocks may help to discover when life first arose on Earth, and analyses of carbon isotope ratios may help to identify climate changes and even to explain major transitions such as the Permian–Triassic extinction event. A relatively recent discipline, molecular phylogenetics, often helps by using comparisons of different modern organisms' DNA and RNA to re-construct evolutionary "family trees"; it has also been used to estimate the dates of important evolutionary developments, although this approach is controversial because of doubts about the reliability of the "molecular clock". Techniques developed in engineering have been used to analyse how ancient organisms might have worked, for example how fast *Tyrannosaurus* could move and how powerful its bite was. Paleontology even contributes to astrobiology, the investigation of possible life on other planets, by developing models of how life may have arisen and by providing techniques for detecting evidence of life.

Subdivisions

As knowledge has increased, paleontology has developed specialised subdivisons. Vertebrate paleontology concentrates on fossils of vertebrates, from the earliest fish to the immediate ancestors of modern mammals. Invertebrate paleontology deals with fossils of invertebrates such as molluscs, arthropods, annelid worms and echinoderms. Paleobotany focuses on the study of fossil plants, but traditionally includes the study of fossil algae and fungi. Palynology, the study of pollen and spores produced by land plants and protists, straddles the border between paleontology and botany, as it deals with both living and fossil organisms. Micropaleontology deals with all microscopic fossil organisms, regardless of the group to which they belong.

Instead of focusing on individual organisms, paleoecology examines the interactions between different organisms, such as their places in food chains, and the two-way interaction between organisms and their environment – for example the development of oxygenic photosynthesis by bacteria hugely increased the productivity and diversity of ecosystems, and also caused the oxygenation of the atmosphere, which in turn was a prerequisite for the evolution of the most complex eucaryotic cells, from which all multicellular organisms are built. Paleoclimatology, although sometimes treated as part of paleoecology, focuses more on the history of Earth's climate and the mechanisms which have changed it – which have sometimes included evolutionary developments, for example the rapid expansion of land plants in the Devonian period removed more carbon dioxide from the atmosphere, reducing the greenhouse effect and thus helping to cause an ice age in the Carboniferous period. Biostratigraphy, the use of fossils to work out the chronological order in which rocks were formed, is useful to both paleontologists and geologists. Biogeography studies the spatial distribution of organisms, and is also linked to geology, which explains how Earth's geography has changed over time.

SOURCES OF EVIDENCE

Body Fossils

Fossils of organisms' bodies are usually the most informative type of evidence. The most common types are wood, bones, and shells. Fossilisation is a rare event, and most fossils are destroyed by erosion or metamorphism before they can be observed. Hence the fossil record is very incomplete, increasingly so further back in time. Despite this, it is often adequate to illustrate the broader patterns of life's history. There are also biases in the fossil record: different environments are more favorable to the preservation of different types of organism or parts of organisms. Further, only the parts of organisms that were already mineralised are usually preserved, such as the shells of molluscs. Since most animal species are soft-bodied, they decay before they can become fossilised. As a result, although there are 30-plus phyla of living animals, two-thirds have never been found as fossils.

Occasionally, unusual environments may preserve soft tissues. These lagerstätten allow paleontologists to examine the internal anatomy of animals that in other sediments are represented only by shells, spines, claws, etc. – if they are preserved at all. However, even lagerstätten present an incomplete picture of life at the time. The majority of organisms living at the time are probably not represented because lagerstätten are restricted to a narrow range of environments, e.g. where soft-bodied organisms can be preserved very quickly by events such as mudslides; and the exceptional events that cause quick burial make it difficult to study the normal environments of the animals. The sparseness of the fossil record means that organisms are expected to exist long before and after they are found in the fossil record – this is known as the Signor-Lipps effect.

Trace Fossils

Trace fossils consist mainly of tracks and burrows, but also include coprolites (fossil feces) and marks left by feeding. Trace fossils are particularly significant because they represent a data source that is not limited to animals with easily-fossilized hard parts, and which reflects organisms' behaviour. Also many traces date from significantly earlier than the body fossils of animals that are thought to have been capable of making them. Whilst exact assignment of trace fossils to their makers is generally impossible, traces may for example provide the earliest physical evidence of the appearance of moderately complex animals (comparable to earthworms).

Geochemical Observations

Geochemical observations may help to deduce the global level of biological activity, or the affinity of a certain fossil. For example geochemical features of rocks may reveal when life first arose on Earth, and may provide evidence of the presence of eucaryotic cells, the type from which all multicellular organisms are built. Analyses of carbon isotope ratios may help to explain major transitions such as the Permian–Triassic extinction event.

Classifying Ancient Organisms

Naming groups of organisms in a way that is clear and widely agreed is important, as some disputes in palaeontology have been based just on misunderstandings over names. Linnean taxonomy is commonly used for classifying living organisms, but runs into difficulties when dealing with newly-discovered organisms that are significantly different from known ones. For example: it is hard to decide at what level to place a new higher-level grouping, e.g. genus or family or order; this is important since the Linnean rules for naming groups are tied to their levels, and hence if a group is moved to a different level it has to be renamed.

Paleontologists generally use approaches based on cladistics, a technique for working out the evolutionary "family tree" of a set of organisms. It works by the logic that, if groups B and C have more similarities to each other than either has to group A, then B and C are more closely related to each other than either is to A. Characters that are compared may be anatomical, such as the presence of a notochord, or molecular, by comparing sequences of DNA or proteins. The result of a successful analysis is a hierarchy of clades – groups that share a common ancestor. Ideally the "family tree" has only two branches leading from each node ("junction"), but sometimes there is too little information to achieve this and paleontologists have to make do with junctions that have several branches. The cladistic technique is sometimes fallible, as some features, such as wings or camera eyes, evolved more than once, convergently – this must be taken into account in analyses.

Evolutionary developmental biology, commonly abbreviated to "Evo Devo", also helps paleontologists to produce "family trees". For example the embryological development of some modern brachiopods suggests that brachiopods may be descendants of the halkieriids, which became extinct in the Cambrian period.

ESTIMATING THE DATES OF ORGANISMS

Paleontology seeks to map out how living things have changed through time. A substantial hurdle to this aim is the difficulty of working out how old fossils are. Beds which preserve fossils typically lack the radioactive elements needed for radiometric dating.

This technique is our only means of giving rocks greater than about 50 million years old an absolute age, and can be accurate to within 0.5% or better. Although radiometric dating requires very careful laboratory work, its basic principle is simple: the rates at which various radioactive elements decay are known, and so the ratio of the radioactive element to the element into which it decays shows how long ago the radioactive element was incorporated into the rock. Radioactive elements are common only in rocks with a volcanic origin, and so the only fossil-bearing rocks that can be dated radiometrically are a few volcanic ash layers.

Consequently, paleontologists must usually rely on stratigraphy to date fossils. Stratigraphy is the science of deciphering the "layer-cake" that is the sedimentary record, and has been compared to a jigsaw puzzle. Rocks normally form relatively horizontal layers, with each layer younger than the one underneath it. If a fossil is found between two layers whose ages are known, the fossil's age must lie between the two known ages. Because rock sequences are not continuous, but may be broken up by faults or periods of erosion, it is very difficult to match up rock beds that are not directly next to one another. However, fossils of species that survived for a relatively short time can be used to link up isolated rocks: this technique is called *biostratigraphy*.

For instance, the conodont *Eoplacognathus pseudoplanus* has a short range in the Middle Ordovician period. If rocks of unknown age are found to have traces of *E. pseudoplanus*, they must have a mid-Ordovician age. Such index fossils must be distinctive, be globally distributed and have a short time range to be useful. However, misleading results are produced if the index fossils turn out to have longer fossil ranges than first thought. Stratigraphy and biostratigraphy can in general provide only relative dating (*A* was before *B*), which is often sufficient for studying evolution. However, this is difficult for some time periods, because of the problems involved in matching up rocks of the same age across different continents.

Family-tree relationships may also help to narrow down the date when lineages first appeared. For instance, if fossils of B or C date to X million years ago and the calculated "family tree" says A was an ancestor of B and C, then A must have evolved more than X million years ago. It is also possible to estimate how long ago two living clades diverged – i.e. approximately how long ago their last common ancestor must have lived – by assuming that DNA mutations accumulate at a constant rate. These "molecular clocks", however, are fallible, and provide only a very approximate timing: for example, they are not sufficiently precise and reliable for estimating when the groups that feature in the Cambrian explosion first evolved, and estimates produced by different techniques may vary by a factor of two.

Overview of the History of Life

The evolutionary history of life stretches back to over 3,000 million years ago, possibly as far as 3,800 million years ago. Earth formed about 4,540 million years ago and, after a collision that formed the Moon about 40 million years later, may have cooled quickly enough to have oceans and an atmosphere about 4,440 million years ago. However there is evidence on the Moon of a Late Heavy Bombardment from 4,000 to 3,800 million years ago. If, as seem likely, such a bombardment struck Earth at the same time, the first atmosphere and oceans may have been stripped away. The oldest clear evidence of life on Earth dates to 3,000 million years ago, although there have been reports, often disputed, of fossil bacteria from 3,400 million years ago and of geochemical evidence for the presence of life 3,800 million years ago. Even the simplest modern organisms are too complex to have emerged directly from non-living materials. Some scientists have proposed that life on Earth was "seeded" from elsewhere, but most research concentrates on various explanations of how life could have arisen independently on Earth.

For about 2,000 million years microbial mats, multi-layered colonies of different types of bacteria, were the dominant life on Earth. The evolution of oxygenic photosynthesis enabled them to play the major role in the oxygenation of the atmosphere from about 2,400 million years ago. This change in the atmosphere increased their effectiveness as nurseries of evolution. While eukaryotes, cells with complex internal structures, may have been present earlier, their evolution speeded up when they acquired the ability to transform oxygen from a poison to a powerful source of energy in their metabolism. This innovation may have come from primitive eukaryotes capturing oxygen-powered bacteria as endosymbionts and transforming them into organelles called mitochondria. The earliest evidence of complex eukaryotes with organelles such as mitochondria, dates from 1,850 million years ago.

Multicellular life is composed only of eukaryotic cells, and the earliest evidence for it is the Francevillian Group Fossils from 2,100 million years ago, although specialization of cells for different functions first appears between 1,430 million years ago (a possible fungus) and 1,200 million years ago (a probable red alga). Sexual reproduction may be a prerequisite for specialization of cells, as an asexual multicellular organism might be at risk of being taken over by rogue cells that retain the ability to reproduce.

The earliest known animals are cnidarians from about 580 million years ago, but these are so modern-looking that the earliest animals must have appeared before then. Early fossils of animals are rare because they did not develop mineralized hard parts that fossilize easily until about 548 million years ago. The earliest modern-looking bilaterian animals appear in the Early Cambrian, along with several "weird wonders" that bear little obvious resemblance to any modern animals. There is a long-running debate about whether this Cambrian explosion was truly a very rapid period of evolutionary experimentation; alternative views are that modern-looking animals began evolving earlier but fossils of their precursors have not yet been found, or that the "weird wonders" are evolutionary "aunts" and "cousins" of modern groups. Vertebrates remained an obscure group until the first fish with jaws appeared in the Late Ordovician.

The spread of life from water to land required organisms to solve several problems, including protection against drying out and supporting themselves against gravity. The earliest evidence of land plants and land invertebrates date back to about 476 million years ago and 490 million years ago respectively. The lineage that produced land vertebrates evolved later but very rapidly between 370 million years ago and 360 million years ago; recent discoveries have overturned earlier ideas about the history and driving forces behind their evolution. Land plants were so successful that they caused an ecological crisis in the Late Devonian, until the evolution and spread of fungi that could digest dead wood.

During the Permian period synapsids, including the ancestors of mammals, may have dominated land environments, but the Permian–Triassic extinction event 251 million years ago came very close to wiping out complex life. During the slow recovery from this catastrophe a previously obscure group, archosaurs, became the most abundant and diverse terrestrial vertebrates. One archosaur group, the dinosaurs, were the dominant land vertebrates for the rest of the Mesozoic, and birds evolved from one group of dinosaurs. During this time mammals' ancestors survived only as small, mainly nocturnal insectivores, but this apparent set-back may have accelerated the development of mammalian traits such as endothermy and hair. After the Cretaceous–Tertiary extinction event 65 million years ago killed off the non-avian dinosaurs – birds are the only surviving dinosaurs – mammals increased rapidly in size and diversity, and some took to the air and the sea.

Fossil evidence indicates that flowering plants appeared and rapidly diversified in the Early Cretaceous, between 130 million years ago and 90 million years ago. Their rapid rise to dominance of terrestrial ecosystems is thought to have been propelled by coevolution with pollinating insects. Social insects appeared around the same time and, although they account for only small parts of the insect "family tree", now form over 50% of the total mass of all insects.

Humans evolved from a lineage of upright-walking apes whose earliest fossils date from over 6 million years ago. Although early members of this lineage had chimp-sized brains, about 25% as big as modern humans', there are signs of a steady increase in brain size after about 3 million years ago. There is a long-running debate about whether *modern* humans are descendants of a single small population in Africa, which then migrated all over the world less than 200,000 years ago and replaced previous hominine species, or arose worldwide at the same time as a result of interbreeding.

Mass Extinctions

Life on earth has suffered occasional mass extinctions at least since 542 million years ago. Although they are disasters at the time, mass extinctions have sometimes accelerated the evolution of life on earth. When dominance of particular ecological niches passes from one group of organisms to another, it is rarely because the new dominant group is "superior" to the old and usually because an extinction event eliminates the old dominant group and makes way for the new one. The fossil record appears to show that the rate of extinction is slowing down, with both the gaps between mass extinctions becoming longer and the average and background rates of extinction decreasing. However, it is not certain whether the actual rate of extinction has altered, since both of these observations could be explained in several ways:

- The oceans may have become more hospitable to life over the last 500 million years and less vulnerable to mass extinctions: dissolved oxygen became more widespread and penetrated to greater depths; the development of life on land reduced the run-off of nutrients and hence the risk of eutrophication and anoxic events; marine ecosystems became more diversified so that food chains were less likely to be disrupted.
- Reasonably complete fossils are very rare, most extinct organisms are represented only by partial fossils, and complete fossils are rarest in the oldest rocks. So paleontologists have mistakenly assigned parts of the same organism to different genera which were often defined solely to accommodate these finds – the story of *Anomalocaris* is an example of this. The risk of this mistake is higher for older fossils because these are often unlike parts of any living organism. Many of the "superfluous" genera are represented by fragments which are not found again and the "superfluous" genera appear to become extinct very quickly.

Biodiversity in the fossil record, which is, "the number of distinct genera alive at any given time; that is, those whose first occurrence predates and whose last occurrence postdates that time" shows a different trend: a fairly swift rise from 542 to 400 million years ago, a slight decline from 400 to 200 million years ago, in which the devastating Permian–Triassic extinction event is an important factor, and a swift rise from 200 million years ago to the present.

History of Paleontology

Although paleontology became established around 1800, earlier thinkers had noticed aspects of the fossil record. The ancient Greek philosopher Xenophanes (570–480 BC) concluded from fossil sea shells that some areas of land were once under water. During the Middle Ages the Persian naturalist Ibn Sina, known as *Avicenna* in Europe. The Chinese naturalist Shen Kuo (1031–1095) proposed a theory of climate change based on the presence of petrified bamboo in regions that in his time were too dry for bamboo. In early modern Europe, the systematic study of fossils emerged as an integral part of the changes in natural philosophy that occurred during the Age of Reason. At the end of the 18th century Georges Cuvier's work established comparative anatomy as a scientific discipline and, by proving that some fossil animals resembled no living ones, demonstrated that animals could become extinct, leading to the emergence of paleontology. The expanding knowledge of the fossil record also played an increasing role in the development of geology, particularly stratigraphy.

The first half of the 19th century saw geological and paleontological activity become increasingly well organized with the growth of geologic societies and museums and an increasing number of professional geologists and fossil specialists. Interest increased for reasons that were not purely scientific, as geology and paleontology helped industrialists to find and exploit natural resources such as coal.

This contributed to a rapid increase in knowledge about the history of life on Earth and to progress in the definition of the geologic time scale, largely based on fossil evidence. In 1822 Henri Marie Ducrotay de Blanville, editor of *Journal de Phisique*, coined the word "paleontology" to refer to the study of ancient living organisms through fossils. As knowledge of life's history continued to improve, it became increasingly obvious that there had been some kind of successive order to the development of life. This encouraged early evolutionary theories on the transmutation of species. After Charles Darwin published *Origin of Species* in 1859, much of the focus of paleontology shifted to understanding evolutionary paths, including human evolution, and evolutionary theory.

The last half of the 19th century saw a tremendous expansion in paleontological activity, especially in North America. The trend continued in the 20th century with additional regions of the Earth being opened to systematic fossil collection.

Fossils found in China near the end of the 20th century have been particularly important as they have provided new information about the earliest evolution of animals, early fish, dinosaurs and the evolution of birds. The last few decades of the 20th century saw a renewed interest in mass extinctions and their role in the evolution of life on Earth. There was also a renewed interest in the Cambrian explosion that apparently saw the development of the body plans of most animal phyla. The discovery of fossils of the Ediacaran biota and developments in paleobiology extended knowledge about the history of life back far before the Cambrian.

Increasing awareness of Gregor Mendel's pioneering work in genetics led first to the development of population genetics and then in the mid-20th century to the modern evolutionary synthesis, which explains evolution as the outcome of events such as mutations and horizontal gene transfer which provide genetic variation, with genetic drift and natural selection driving changes in this variation over time. Within the next few years the role and operation of DNA in genetic inheritance were discovered, leading to what is now known as the "Central Dogma" of molecular biology. In the 1960s molecular phylogenetics, the investigation of evolutionary "family trees" by techniques derived from biochemistry, began to make an impact, particularly when it was proposed that the human lineage had diverged from apes much more recently than was generally thought at the time. Although this early study compared proteins from apes and humans, most molecular phylogenetics research is now based on comparisons of RNA and DNA.

PALYNOLOGY

Palynology is the science that studies contemporary and fossil palynomorphs, including pollen, spores, orbicules, dinoflagellate cysts, acritarchs, chitinozoans and scolecodonts, together with particulate organic matter (POM) and kerogen found in sedimentary rocks and sediments. Palynology does not include diatoms, foraminiferans or other organisms with silicaceous or calcareous exoskeletons. Palynology is an interdisciplinary science and is a branch of earth science (geology or geological science) and biological science (biology), particularly plant science (botany). Stratigraphical palynology is a branch of micropalaeontology and paleobotany which studies fossil palynomorphs from the Precambrian to the Holocene.

A HISTORY OF PALYNOLOGY

Early History

The earliest reported observations of pollen under a microscope are likely to have been in the 1640s by the English botanist Nehemiah Grew who described pollen, the stamen and successfully predicted that pollen was required for successful reproduction in plants. As microscopes began to improve further studies included work by Robert Kidston and P. Reinsch examined the presence of spores in coal and compared them to modern spores. The early pioneers also included Christian Gottfried Ehrenberg (radiolarians and diatoms), Gideon Mantell (desmids) and Henry Hopley White (dinoflagellates).

Modern Palynology

The earliest quantitative analysis of pollen was published by Lennart von Post who laid out the foundations of modern pollen analysis in his Kristiania lecture of 1916. Pollen analysis was initially confined to Nordic countries because many early publications were in Nordic languages. This isolation ended with the publication of Gunnar Erdtman's thesis of 1921 when pollen analysis became widespread throughout Europe and North America for use in studies of Quaternary vegetation and climate change. The term *palynology* was introduced by Hyde and Williams in 1944, following correspondence with the Swedish geologist Antevs, in the pages of the Pollen Analysis Circular (one of the first journals devoted to pollen analysis, produced by Paul Sears in North America). Hyde and Williams chose *palynology* on the basis of the Greek words *paluno* meaning 'to sprinkle' and *pale* meaning 'dust' (and thus similar to the Latin word *pollen*).

Methods of Study

Palynomorphs are broadly defined as organic-walled microfossils between 5 and 500 micrometres in size. They are extracted from rocks and sediment cores both physically, by wet sieving, often after ultrasonic treatment, and chemically, by using chemical digestion to remove the non-organic fraction.

Chemical Preparation

Chemical digestion follows a number of steps. Initially the only chemical treatment used by researchers was treatment with KOH to remove humic substances; defloculation was accomplished through surface treatment or ultra-sonic treatment, although sonification may cause the pollen exine to rupture.

The use of hydrofluoric acid (HF) to digest silicate minerals was introduced by Assarson and Granlund in 1924, greatly reducing the amount of time required to scan slides for palynomorphs. Palynological studies using peats presented a particular challenge because of the presence of well preserved organic material including fine rootlets, moss leaflets and organic litter. This was the last major challenge in the chemical preparation of materials for palynological study. Acetolysis was developed by Gunnar Erdtman and his brother to remove these fine cellulose materials by dissolving them.. In acetolysis the material is treated with acetic anhydride and sulfuric acid, dissolving cellulistic materials and providing better visibility for palynomorphs. Some steps of the chemical treatments require special care for safety reason, in particular the use of HF which diffuses very fast through the skin and could cause severe chemical burns.

Other treatment include kerosene flotation for chitinous materials.

Analysis

Once samples have been prepared chemically, they are mounted on microscope slides using silicon oil, glycerol or glycerol-jelly and examined using light microscopy or scanning electron microscopy. Researchers will often study either modern samples from a number of unique sites within a given area, or samples from a single site with a record through time, such as samples obtained from peat or lake sediments. More recent studies have used the modern analog technique in which paleo-samples are compared to modern samples for which the parent vegetation is known

When the slides are observed under a microscope the researcher will count the number of grains from each pollen taxon. This record is then used to produce a pollen diagram. This data can be used to detect anthropogenic effects such as logging, traditional patterns of land use or long term changes in regional climate Palynology can be applied to problems in many fields including geology, botany, paleontology, archaeology, pedology (soil study), and geography.

Applications

Palynology is used for a diverse range of applications, related to many scientific disciplines:

- Biostratigraphy and geochronology. Geologists use palynological studies in biostratigraphy to correlate strata and determine the relative age of a given bed, horizon, formation or stratigraphical sequence.
- Palaeoecology and climate change. Palynology can be used to reconstruct past vegetation (land plants) and marine and freshwater phytoplankton communities, and so infer past environmental (palaeoenvironmental) and palaeoclimatic conditions.
- Organic palynofacies studies, which examine the preservation of the particulate organic matter and palynomorphs provides information on the depositional environment of sediments and depositional palaeoenvironments of sedimentary rocks.
- Geothermal alteration studies examine the colour of palynomorphs extracted from rocks to give the thermal

alteration and maturation of sedimentary sequences, which provides estimates of maximum palaeotemperatures.

- Limnology studies. Freshwater palynomorphs and animal and plant fragments, including the prasinophytes and desmids (green algae) can be used to study past lake levels and long term climate change.
- Taxonomy and evolutionary studies.
- Forensic palynology — the study of pollen and other palynomorphs for evidence at a crime scene.
- Allergy studies. Studies of the geographic distribution and seasonal production of pollen, can help sufferers of allergies such as hay fever.
- Melissopalynology — the study of pollen and spores found in honey.
- Archaeological palynology examines human uses of plants in the past. This can help determine seasonality of site occupation, presence or absence of agricultural practices or products and plant-related activity areas within an archaeological context. Bonfire Shelter is one such example of this application.

Because the distribution of acritarchs, chitinozoans, dinoflagellate cysts, pollen and spores provides evidence of stratigraphical correlation through biostratigraphy and palaeoenvironmental reconstruction, one common and lucrative application of palynology is in oil and gas exploration. Palynology also allows scientists to infer the climatic conditions from the vegetation present in an area thousands or millions of years ago. This is a fundamental part of research into climate change.

MICROPALEONTOLOGY

Micropaleontology (also sometimes spelled as micropalaeontology) is that branch of paleontology which studies microfossils.

Microfossils

Microfossils are fossils generally not larger than four millimeters, and commonly smaller than one millimeter, the study of which requires the use of light or electron microscopy. Fossils which can be studied with the naked eye or low-powered magnification, such as a hand lens, are referred to as macrofossils. Obviously, it can be hard to decide whether or not some organisms should be considered microfossils, and so there is no fixed size boundary. For example, some colonial organisms, such as bryozoa (especially the Cheilostomata) have relatively large colonies, but are classified on the basis of fine skeletal details of the tiny individuals of the colony. Most bryozoan specialists tend to consider themselves paleontologists, rather than micropaleontologists, but many micropaleontologists also study bryozoa.

In another example, many fossil genera of Foraminifera, which are protists, are known from shells (called "tests") that were as big as coins, such as the genus *Nummulites*. Microfossils are a common feature of the geological record, from the Precambrian to the Holocene. They are most common in deposits of marine environments, but also occur in brackish water, fresh water and terrestrial sedimentary deposits. While every kingdom of life is represented in the microfossil record, the most abundant forms are protist skeletons or cysts from the Chrysophyta, Pyrrhophyta, Sarcodina, acritarchs and chitinozoans, together with pollen and spores from the vascular plants.

Areas of Study

Micropaleontology can be roughly divided into four areas of study on the basis of microfossil composition: (a) calcareous, as in coccoliths and foraminifera, (b) phosphatic, as in the study of some vertebrates, (c) siliceous, as in diatoms and radiolaria, or (d) organic, as in the pollen and spores studied in palynology. This division reflects differences in the mineralogical and chemical composition of microfossil remains (and therefore in the methods of fossil recovery) rather than any strict taxonomic or ecological distinctions. Most researchers in this field, known as micropaleontologists, are typically specialists in one or more taxonomic groups.

Calcareous Microfossils

Calcareous [CaCO3] microfossils include Coccoliths, Foraminifera, Calcareous dinoflagellates, and Ostracods (seed shrimp).

Phosphatic Microfossils

Phosphatic microfossils include Conodonts (tiny oral structures of an extinct chordate group), some scolecodonts ("worm" jaws), Shark spines and teeth, and other Fish remains (collectively called "ichthyoliths").

Siliceous Microfossils

Siliceous microfossils include Diatoms, Radiolaria, Silicoflagellates, phytoliths, some scolecodonts ("worm" jaws), and spicules.

Organic Microfossils

The study of organic microfossils is called palynology. Organic microfossils include pollen, spores, Chitinozoans (thought to be the egg cases of marine invertebrates), Scolecodonts ("worm" jaws), Acritarchs, Dinoflagellate cysts, and fungal remains.

Methods

Sediment or rock samples are collected from either cores or outcrops, and the microfossils they contain extracted by a variety of physical and chemical laboratory techniques, including sieving, density separation by centrifuge, and chemical digestion of the unwanted fraction. The resulting concentrated sample of microfossils is then mounted on a slide for analysis, usually by light microscope. Taxa are then identified and counted. The very large numbers of microfossils that a small sediment sample can often yield allows the collection of statistically robust datasets which can be subjected to multivariate analysis. A typical microfossil study will involve identification of a few hundred specimens from each of ten to a hundred samples.

Applications of Micropaleontology

Microfossils are especially noteworthy for their importance in biostratigraphy. Since microfossils are often extremely abundant, widespread, and quick to appear and disappear from the stratigraphic record, they constitute ideal index fossils from a biostratigraphic perspective. In addition, the planktonic and nektonic habits of some microfossils gives them the added bonus of appearing across a wide range of facies or paleoenvironments, as well as having near-global distribution, making biostratigraphic correlation even more powerful and effective.

Microfossils also provide some of the most important records of global environmental change on long-timescales, particularly from deep-sea sediments. Across vast areas of the ocean floor, the shells of planktonic micro-organisms sinking from surface waters provide the dominant source of sediment, and they continuously accumulate (typically at rates of 20-50 million per million years). Study of changes in assemblages of microfossils and of changes in their shell chemistry (e.g., oxygen isotope composition) are fundamental to research on climate change in the geological past.

In addition to providing an excellent tool for sedimentary rock-body dating and for paleoenvironmental reconstruction – heavily used in both petroleum geology and paleoceanography – micropaleontology has also found a number of less orthodox applications, such as its growing role in forensic police investigation or in determining the provenance of archaeological artefacts.

GEOMICROBIOLOGY

Geomicrobiology is the result of the combination of geology and microbiology. The field of geomicrobiology concerns the role of microbe and microbial processes in geological and geochemical processes and vice-versa. The field is especially important when dealing with microorganisms in aquifers and public drinking water supplies. Another area of investigation in geomicrobiology is the study of extremophile organisms, the microorganisms that thrive in environments normally considered hostile. Such environments may include extremely hot (hot springs or mid-ocean ridge black smoker) environments, extremely saline environments, or even space environments such as Martian soil or comets.

Recent observations and research in hyper-saline lagoon environments in Brazil and Australia have shown that anaerobic sulfate-reducing bacteria may be directly involved in the formation of dolomite. This suggests the alteration and replacement of limestone sediments by dolomitization in ancient rocks was possibly aided by ancestors to these anaerobic bacteria. Some bacteria use metal ions as their energy source. They convert (or chemically reduce) the dissolved metal ions from one electrical state to another. This reduction releases energy for the bacteria's use, and, as a side product, serves to concentrate the metals into what ultimately become *ore deposits*. Certain iron, uranium and even gold ores are thought to have formed as the result of microbe action.

Microbes are being studied and used to degrade organic and even nuclear waste pollution and assist in environmental cleanup. (It should be stressed however that microorganisms can not decrease the total radioactivity of nuclear waste.) An application of geomicrobiology is bioleaching, the use of microbes to extract metals from mine waste.

TECHNIQUES USED IN GEOARCHAEOLOGY

Column Sampling

Column sampling is a technique of collecting samples from a section for analyzing and detecting the buried processes down the profile of the section. Narrow metal tins are bashed into the section in a series to collect the complete profile for study. If more than one tin is needed they are arranged offset and overlapping to one side so the complete profile can be rebuilt offsite in laboratory conditions.

Loss on Ignition Testing

Loss on ignition testing for soil organic content.-a technique of measuring organic content in soil samples. Samples taken from a known place in the profile collected by column sampling are weighed then placed in a fierce oven which burns off the organic content.

The resulting cooked sample is weighed again and the resulting loss in weight is an indicator of organic content in the profile at a certain depth. These readings are often used to detect buried soil horizons. A buried soil's horizons may not be visible in section and this horizon is an indicator of possible occupation levels. Ancient land surfaces especially from the prehistoric era can be difficult to discern so this technique is useful for evaluating an areas potential for prehistoric surfaces and archaeological evidence. Comparative measurements down the profile are made and a sudden rise in organic content at some point in the profile combined with other indicators is strong evidence for buried surfaces.

Magnetic Susceptibility Analysis

The magnetic susceptibility of a material is a measure of its ability to become magnetised by an external magnetic field (Dearing, 1999). The magnetic susceptibility of a soil reflects the presence of magnetic iron-oxide minerals such as maghaematite; just because a soil contains a lot of iron does not mean that it will have high magnetic susceptibility. Magnetic forms of iron can be formed by burning and microbial activity such as occurs in top soils and some anaerobic deposits. Magnetic iron compounds can also be found in igneous and metamorphic rocks.

The relationship between iron and burning means that magnetic susceptibility is often used for:

- Site prospection, to identify areas of archaeological potential prior to excavation.
- Identifying hearth areas and the presence of burning residues in deposits (Tite and Mullins, 1971).
- Explaining whether areas of reddening are due to burning or other natural processes such as gleying (waterlogging).

The relationship between soil formation and magnetic susceptibility means that it can also be used to:

- Identify buried soils in depositional sequences.
- Identify redeposited soil materials in peat, lake sediments etc.

Phosphate and Orthophosphate Content with Spectrophotometetry

Phosphate in man-made soils derives from people, their animals, rubbish and bones. 100 people excrete about 62kg of phosphate annually, with about the same from their rubbish. Their animals excrete even more. A human body contains about 650g of PO4,(500g-80% in the skeleton), which results in elevated levels in burial sites. Most is quickly immobilised on the clay of the soil and 'fixed', where it can persist for thousands of years. For a 1 ha site this corresponds to about

150kg PO4 ha-1yr-1 about 0.5% to 10% of that already present in most soils. Therefore it doesn't take long for human occupation to make orders of magnitude differences to the phosphate concentration in soil. Phosphorus exist in different 'pools' in the soil 1) organic (available), 2) occluded (adsorbed), 3) bound (chemically bound). Each of these pools can be extracted using progressively more aggressive chemicals. Some workers (Eidt especially), think that the ratios between these pools can give information about past land use, and perhaps even dating. Whatever the method of getting the phosphorus from the soil into solution, the method of detecting it is usually the same. This uses the 'molybdate blue' reaction, where the depth of the colour is proportional to phosphorus concentration. In the lab, this is measured using a colorimeter, where light shining through a standard cell produces an electrical current proportional to the light attenuation. In the field, the same reaction is used on detector sticks, which are compared to a colour chart. Phosphate concentrations can be plotted on archaeological plans to show former activity areas, and is also used to prospect for sites in the wider landscape.

Particle Size Analysis

The particle size distribution of a soil sample may indicate the conditions under which the strata or sediment were deposited. Particle sizes are generally separated by means of dry or wet sieving (coarse samples such as till, gravel and sands, sometimes coarser silts) or by measuring the changes of the density of a dispersed solution (in sodiumpyrophosphate, for example))of the sample (finer silts, clays). A rotating clock-glass with a very fine-grained dispersed sample under a heat lamp is useful in separating particles. The results are plotted on curves which can be analyzed with statistical methods for particle distribution and other parameters. The fractions received can be further investigated for cultural indicators, macro-and microfossils and other interesting features, so particle size analysis is in fact the first thing to do when handling these samples.

Trace Element Geochemistry

Trace element geochemistry is the study of the abundances of elements in geological materials that do not occur in a large quantity in these materials. Because these trace elements concentration are determined by a large number of particular situations under which a certain geological material is formed, they are usually unique between two locations which contain the same type of rock or other geological material. Geoarchaeologists use this uniqueness in trace element geochemistry to trace ancient patterns of resource-acquisition and trade.

For example, researchers can look at the trace element composition of obsidian artifacts in order to "fingerprint" those artifacts. They can then study the trace element composition of obsidian outcrops in order to determine the original source of the raw material used to make the artifact.

Clay Mineralogy Analysis

Geoarchaeologists study the mineralogical characteristics of pots through macroscopic and microscopic analyses. They can use these characteristics to understand the various manufacturing techniques used to make the pots, and through this, to know which production centers likely made these pots. They can also use the mineralogy to trace the raw materials used to make the pots to specific clay deposits.

Ostracode Analysis

Naturally occurring Ostracods in freshwater bodies are impacted by changes in salinity and pH due to human activities. Analysis of Ostracod shells in sediment columns show the changes brought about by farming and habitation activities. This record can be correlated with age dating techniques to help identify changes in human habitation patterns and population migrations.

Archaeological Geology

Archaeological geology (term coined by Werner Kasig, 1980), is a sub-field of geology which emphasises the value of earth constituents for human life. Over the last decades, archaeologists and historians have faced the necessity to reconstruct ancient settlement history not only through the study of the material excavated, but also with the use of palaeo-environmental parameters.

BIOSPHERE RESERVES OF INDIA

The Indian government has established 15 Biosphere Reserves of India, (categories roughly corresponding to IUCN Category V Protected areas), which protect larger areas of natural habitat (than a National Park or Animal Sanctuary), and often include one or more National Parks and/or preserves, along buffer zones that are open to some economic uses. Protection is granted not only to the flora and fauna of the protected region, but also to the human communities who inhabit these regions, and their ways of life. Seven of the fifteen biosphere reserves are a part of the World Network of Biosphere Reserves, based on the UNESCO Man and the Biosphere (MAB) Programme list.

- Nilgiri Biosphere Reserve
- Nanda Devi Biosphere Reserve
- Sundarbans national park
- Gulf of Mannar Biosphere Reserve
- Pachmarhi Biosphere Reserve
- Nokrek Biosphere Reserve
- Simlipal Biosphere Reserve.

The last three are among the 22 new sites recently added by UNESCO to its World Network of Biosphere Reserves. Their inclusion raises the number of Indian sites on the World list to seven, from a total of 15 Biosphere Reserves across the subcontinent. Recently India has added one more area as bsr namely cold desert of Himachal Pradesh taking the total count to 16.

5

The Hydrosphere

A hydrosphere in physical geography describes the combined mass of water found on, under, and over the surface of a planet. The total mass of the Earth's hydrosphere is about 1.4×10^{18} tonnes, which is about 0.023% of the Earth's total mass. About 20×10^{12} tonnes of this is in the Earth's atmosphere (the volume of one tonne of water is approximately 1 cubic metre). Approximately 71% of the Earth's surface, an area of some 361 million square kilometres (139.5 million square miles), is covered by ocean. The average salinity of the Earth's oceans is about 35 grams of salt per kilogram of sea water (35 ‰).

Other Hydrospheres

A thick hydrosphere is thought to exist around the Jovian moon Europa. The outer layer of this hydrosphere is almost entirely ice, but current models predict that there is an ocean up to 100 km in depth underneath the ice. This ocean remains in a liquid form because of tidal flexing of the moon in its orbit around Jupiter. The volume of Europa's hydrosphere is 3×10^{18} m^3, 2.3 times that of Earth. It has been suggested that the Jovian moon Ganymede and the Saturnian moon Enceladus may also possess sub-surface oceans. However the ice covering is expected to be thicker on Jupiter's Ganymede than on Europa.

Hydrological Cycle

Insolation, or energy (in the form of heat and light) from the sun, provides the energy necessary to cause evaporation from all wet surfaces including oceans, rivers, lakes, soil and the leaves of plants. Water vapour is further released as transpiration from vegetation and from humans and other animals.

HYDROLOGY

Hydrology is the study of the movement, distribution, and quality of water throughout the Earth, including the hydrologic cycle, water resources and environmental watershed sustainability. A practitioner of hydrology is a hydrologist, working within the fields of either earth or environmental science, physical geography, geology or civil and environmental engineering. Domains of hydrology include hydrometeorology, surface hydrology, hydrogeology, drainage basin management and water chemistry, where water plays the central role. Oceanography and meteorology are not included because water is only one of many important aspects. Hydrological research can inform environmental engineering, policy and planning.

History of Hydrology

Hydrology has been a subject of investigation and engineering for millennia. For example, about 4000 B.C. the Nile was dammed to improve agricultural productivity of previously barren lands. Mesopotamian towns were protected from flooding with high earthen walls. Aqueducts were built by the Greeks and Ancient Romans, while the History of China shows they built irrigation and flood control works.

The ancient Sinhalese used hydrology to build complex irrigation Works in Sri Lanka, also known for invention of the Valve Pit which allowed construction of large reservoirs, anicuts and canals which still function. Marcus Vitruvius, in the first century B.C., described a philosophical theory of the hydrologic cycle, in which precipitation falling in the mountains infiltrated the Earth's surface and led to streams and springs in the lowlands. With adoption of a more scientific approach, Leonardo da Vinci and Bernard Palissy independently reached an accurate representation of the hydrologic cycle. It was not until the 17th century that hydrologic variables began to be quantified.

Pioneers of the modern science of hydrology include Pierre Perrault, Edme Mariotte and Edmund Halley. By measuring rainfall, runoff, and drainage area, Perrault showed that rainfall was sufficient to account for flow of the Seine. Marriotte combined velocity and river cross-section measurements to obtain discharge, again in the Seine. Halley showed that the evaporation from the Mediterranean Sea was sufficient to account for the outflow of rivers flowing into the sea. Advances in the 18th century included the Bernoulli piezometer and Bernoulli's equation, by Daniel Bernoulli, the Pitot tube. The 19th century saw development in groundwater hydrology, including Darcy's law, the Dupuit-Thiem well formula, and Hagen-Poiseuille's capillary flow equation.

Rational analyses began to replace empiricism in the 20th century, while governmental agencies began their own hydrological research programs. Of particular importance were Leroy Sherman's unit hydrograph, the infiltration theory of Robert E. Horton, and C.V. Theis's Aquifer test/equation describing well hydraulics. Since the 1950s, hydrology has been approached with a more theoretical basis than in the past, facilitated by advances in the physical understanding of hydrological processes and by the advent of computers and especially Geographic Information Systems (GIS).

Hydrologic Cycle

The central theme of hydrology is that water circulates throughout the Earth through different pathways and at different rates. The most vivid image of this is in the evaporation of water from the ocean, which forms clouds. These clouds drift over the land and produce rain. The rainwater flows into lakes, rivers, or aquifers. The water in lakes, rivers, and aquifers then either evaporates back to the atmosphere or eventually flows back to the ocean, completing a cycle. Water changes its state of being several times throughout this cycle

Branches of Hydrology

Chemical hydrology is the study of the chemical characteristics of water. Ecohydrology is the study of interactions between organisms and the hydrologic cycle. Hydrogeology is the study of the presence and movement of ground water. Hydroinformatics is the adaptation of information technology to hydrology and water resources applications. Hydrometeorology is the study of the transfer of water and energy between land and water body surfaces and the lower atmosphere. Isotope hydrology is the study of the isotopic signatures of water. Surface hydrology is the study of hydrologic processes that operate at or near Earth's surface.

Hydrologic Measurements

Measurement is fundamental for assessing water resources and understanding the processes involved in the hydrologic cycle. Because the hydrologic cycle is so diverse, hydrologic measurement methods span many disciplines: including soils, oceanography, atmospheric science, geology, geophysics and limnology, to name a few. Here, hydrologic measurement methods are organized by hydrologic sub-disciplines.

Quantifying groundwater flow and transport

- Aquifer characterization
  - * Flow direction:
    - — Piezometer-groundwater pressure and, by inference, groundwater depth
    - — Conductivity, storativity, transmisivity
    - — Geophysical methods:
- Vadose zone characterization
  - * Infiltration:
    - — Infiltrometer-infiltration
  - * Soil moisture:
    - — Capacitance probe-soil moisture
    - — Time domain reflectometer-soil moisture
    - — Tensiometer-soil moisture
    - — Solute sampling
    - — Geophysical methods.

Quantifying surface water flow and transport:

- Direct and indirect discharge measurements
  - * Stream gauge-stream flow
  - * Tracer techniques
  - * Chemical transport
  - * Sediment transport and erosion
  - * Stream-aquifer exchange.

Quantifying exchanges at the land-atmospheric boundary:

- Precipitation

* Bulk rain events:
  - Disdrometer-precipitation characteristics
  - Radar-cloud properties, rain rate estimation, hail and snow detection
  - Rain gauge-rain and snowfall
  - Satellite-rainy area identification, rain rate estimation, land-cover/land-use, soil moisture
  - Sling psychrometer-humidity.
* Snow, hail and ice
* Dew, mist and fog

- Evaporation
  * from water surfaces
  * Evaporation-Symon's evaporation pan
  * from plant surfaces
  * through the boundary layer
- Transpiration
  * Natural ecosystems
  * Agronomic ecosystems
- Momentum
- Heat flux
  * Energy budgets

Uncertainty Analyses

Remote sensing of hydrologic processes:

- Land based sensors
- Airborne Sensors
- Satellite sensors.

Water quality

- Sample collection
- In-situ methods
- Physical measurements (includes sediment concentration)
- Collection of samples to quantify Organic Compounds
- Collection of samples to quantify Inorganic Compounds
- Analysis of aqueous Organic Compounds

- Analysis of aqueous Inorganic Compounds
- Microbiological sampling and analysis.

Integrating measurement and modeling:

- Budget analyses
- Parameter estimation
- Scaling in time and space
- Data assimilation
- Quality control of data.

Hydrologic Prediction

Observations of hydrologic processes are used to make predictions of the future behaviour of hydrologic systems (water flow, water quality). One of the major current concerns in hydrologic research is the Prediction in Ungauged Basins (PUB), i.e. in basins where no or only very few data exist.

Statistical Hydrology

By analysing the statistical properties of hydrologic records, such as rainfall or river flow, hydrologists can estimate future hydrologic phenomena, assuming the characteristics of the processes remain unchanged. These estimates are important for engineers and economists so that proper risk analysis can be performed to influence investment decisions in future infrastructure and to determine the yield reliability characteristics of water supply systems. Statistical information is utilised to formulate operating rules for large dams forming part of systems which include agricultural, industrial and residential demands.

Hydrologic Modeling

Hydrologic models are simplified, conceptual representations of a part of the hydrologic cycle. They are primarily used for hydrologic prediction and for understanding hydrologic processes. Two major types of hydrologic models can be distinguished:

- *Models based on data.* These models are black box systems, using mathematical and statistical concepts to link a certain input (for instance rainfall) to the model output (for instance runoff). Commonly used techniques are regression, transfer functions, and system identification. The simplest of these models may be linear models, but it is common to deploy non-linear components to represent some general aspects of a catchment's response without going deeply into the real physical processes involved. An example of such an aspect is the well-known behaviour that a catchment will respond much more quickly and strongly when it is already wet than when it is dry..

- *Models based on process descriptions.* These models try to represent the physical processes observed in the real world. Typically, such models contain representations of surface runoff, subsurface flow, evapotranspiration, and channel flow, but they can be far more complicated. These models are known as deterministic hydrology models. Deterministic hydrology models can be subdivided into single-event models and continuous simulation models.

Recent research in hydrologic modeling tries to have a more global approach to the understanding of the behaviour of hydrologic systems to make better predictions and to face the major challenges in water resources management.

Hydrologic Transport

Water movement is a significant means by which other material, such as soil or pollutants, are transported from place to place. Initial input to receiving waters may arise from a point source discharge or a line source or area source, such as surface runoff. Since the 1960s rather complex mathematical models have been developed, facilitated by the availability of high speed computers. The most common pollutant classes analyzed are nutrients, pesticides, total dissolved solids and sediment.

Applications of Hydrology

- Determining the water balance of a region.
- Determining the agricultural water balance.
- Designing riparian restoration projects.
- Mitigating and predicting flood, landslide and drought risk.

- Real-time flood forecasting and flood warning.
- Designing irrigation schemes and managing agricultural productivity.
- Part of the hazard module in catastrophe modeling.
- Providing drinking water.
- Designing dams for water supply or hydroelectric power generation.
- Designing bridges.
- Designing sewers and urban drainage system.
- Analyzing the impacts of antecedent moisture on sanitary sewer systems.
- Predicting geomorphological changes, such as erosion or sedimentation.
- Assessing the impacts of natural and anthropogenic environmental change on water resources.
- Assessing contaminant transport risk and establishing environmental policy guidelines.

LIMNOLOGY

Limnology is the study of inland waters. It is often regarded as a division of ecology or environmental science. It covers the biological, chemical, physical, geological, and other attributes of all inland waters (running and standing waters, both fresh and saline, natural or man-made). This includes the study of lakes and ponds, rivers, springs, streams and wetlands. A more recent sub-discipline of limnology, termed landscape limnology, studies, manages, and conserves these aquatic ecosystems using a landscape perspective. Limnology is closely related to aquatic ecology and hydrobiology, which study aquatic organisms in particular regard to their hydrological environment.

History

The term limnology was coined by François-Alphonse Forel (1841–1912) who established the field with his studies of Lake Geneva. Interest in the discipline rapidly expanded, and in 1922 August Thienemann (a German zoologist) and Einar Naumann (a Swedish botanist) co-founded the International Society of Limnology (SIL, for originally Societas Internationalis Limnologiae). Forel's original definition of limnology, "the oceanography of lakes", was expanded to encompass the study of all inland waters.

HYDROGEOLOGY

Hydrogeology (*hydro*-meaning water, and-*geology* meaning the study of the Earth) is the area of geology that deals with the distribution and movement of groundwater in the soil and rocks of the Earth's crust, (commonly in aquifers). The term geohydrology is often used interchangeably. Some make the minor distinction between a hydrologist or engineer applying themselves to geology (geohydrology), and a geologist applying themselves to hydrology (hydrogeology).

Hydrogeology is an interdisciplinary subject; it can be difficult to account fully for the chemical, physical, biological and even legal interactions between soil, water, nature and society. The study of the interaction between groundwater movement and geology can be quite complex. Groundwater does not always flow in the subsurface down-hill following the surface topography; groundwater follows pressure gradients (flow from high pressure to low) often following fractures and conduits in circuitous paths. Taking into account the interplay of the different facets of a multi-component system often requires knowledge in several diverse fields at both the experimental and theoretical levels. The following is a more traditional introduction to the methods and nomenclature of saturated subsurface hydrology, or simply hydrogeology.

Hydrogeology in Relation to Other Fields

Hydrogeology, as stated above, is a branch of the earth sciences dealing with the flow of water through aquifers and other shallow porous media (typically less than 450 m or 1,500 ft below the land surface.) The very shallow flow of water in the subsurface (the upper 3 m or 10 ft) is pertinent to the fields of soil science, agriculture and civil engineering, as well as to hydrogeology. The general flow of fluids (water, hydrocarbons, geothermal fluids, etc.) in deeper formations is also a concern of geologists, geophysicists and petroleum geologists. Groundwater is a slow-moving, viscous fluid (with a Reynolds number less than unity); many of the empirically derived laws of groundwater flow can be alternately derived in fluid mechanics from the special case of Stokes flow (viscosity and pressure terms, but no inertial term). The mathematical relationships used to describe the flow of water through porous media are the diffusion and Laplace equations, which have applications in many diverse fields. Steady groundwater flow (Laplace equation) has been simulated using electrical, elastic and heat conduction analogies. Transient groundwater flow is analogous to the diffusion of heat in a solid, therefore some solutions to hydrological problems have been adapted from heat transfer literature.

Traditionally, the movement of groundwater has been studied separately from surface water, climatology, and even the chemical and microbiological aspects of hydrogeology (the processes are uncoupled). As the field of hydrogeology matures, the strong interactions between groundwater, surface water, water chemistry, soil moisture and even climate are becoming more clear.

Definitions and Material Properties

One of the main tasks a hydrogeologist typically performs is the prediction of future behavior of an aquifer system, based on analysis of past and present observations. Some hypothetical, but characteristic questions asked would be:

- Can the aquifer support another subdivision?
- Will the river dry up if the farmer doubles his irrigation?
- Did the chemicals from the dry cleaning facility travel through the aquifer to my well and make me sick?
- Will the plume of effluent leaving my neighbour's septic system flow to my drinking water well?

Most of these questions can be addressed through simulation of the hydrologic system (using numerical models or analytic equations). Accurate simulation of the aquifer system requires knowledge of the aquifer properties and boundary conditions. Therefore a common task of the hydrogeologist is determining aquifer properties using aquifer tests.

In order to further characterize aquifers and aquitards some primary and derived physical properties are introduced below. Aquifers are broadly classified as being either confined or unconfined (water table aquifers), and either saturated or unsaturated; the type of aquifer affects what properties control the flow of water in that medium (e.g., the release of water from storage for confined aquifers is related to the storativity, while it is related to the specific yield for unconfined aquifers).

Hydraulic Head

Changes in hydraulic head (h) are the driving force which causes water to move from one place to another. It is composed of pressure head (ψ) and elevation head (z). The head gradient is the change in hydraulic head per length of flowpath, and appears in Darcy's law as being proportional to the discharge.

Hydraulic head is a directly measurable property that can take on any value (because of the arbitrary datum involved in the z term); ψ can be measured with a pressure transducer (this value can be negative, e.g., suction, but is positive in saturated aquifers), and z can be measured relative to a surveyed datum (typically the top of the well casing). Commonly, in wells tapping unconfined aquifers the water level in a well is used as a proxy for hydraulic head, assuming there is no vertical gradient of pressure. Often only *changes* in hydraulic head through time are needed, so the constant elevation head term can be left out ($\Delta h = \Delta\psi$). A record of hydraulic head through time at a well is a hydrograph or, the changes in hydraulic head recorded during the pumping of a well in a test are called drawdown.

Porosity

Porosity (n) is a directly measurable aquifer property; it is a fraction between 0 and 1 indicating the amount of pore space between unconsolidated soil particles or within a fractured rock. Typically, the majority of groundwater (and anything dissolved in it) moves through the porosity available to flow (sometimes called effective porosity). Permeability is an expression of the connectedness of the pores. For instance, an unfractured rock unit may have a high *porosity* (it has lots of *holes* between its constituent grains), but a low *permeability* (none of the pores are connected). An example of this phenomenon is pumice, which, when in its unfractured state, can make a poor aquifer.

Porosity does not directly affect the distribution of hydraulic head in an aquifer, but it has a very strong effect on the migration of dissolved contaminants, since it affects groundwater flow velocities through an inversely proportional relationship.

Water Content

Water content (θ) is also a directly measurable property; it is the fraction of the total rock which is filled with liquid water. This is also a fraction between 0 and 1, but it must also be less than or equal to the total porosity.

The water content is very important in vadose zone hydrology, where the hydraulic conductivity is a strongly nonlinear function of water content; this complicates the solution of the unsaturated groundwater flow equation.

Hydraulic Conductivity

Hydraulic conductivity (K) and transmissivity (T) are indirect aquifer properties (they cannot be measured directly). T is the K integrated over the vertical thickness (b) of the aquifer ($T=Kb$ when K is constant over the entire thickness).

These properties are measures of an aquifer's ability to transmit water. Intrinsic permeability ($\hat{e}$) is a secondary medium property which does not depend on the viscosity and density of the fluid (K and T are specific to water); it is used more in the petroleum industry.

Specific Storage and Specific Yield

Specific storage (S_s) and its depth-integrated equivalent, storativity ($S=S_sb$), are indirect aquifer properties (they cannot be measured directly); they indicate the amount of groundwater released from storage due to a unit depressurization of a confined aquifer. They are fractions between 0 and 1.

Specific yield (S_y) is also a ratio between 0 and 1 (S_y d– porosity) and indicates the amount of water released due to drainage from lowering the water table in an unconfined aquifer.

The value for specific yield is less than the value for porosity because some water will remain in the medium even after drainage due to molecular forces. Often the porosity or effective porosity is used as an upper bound to the specific yield. Typically S_y is orders of magnitude larger than S_s.

Contaminant Transport Properties

Often we are interested in how the moving groundwater water will move dissolved contaminants around (the sub-field of contaminant hydrogeology). The contaminants can be man-made (e.g., petroleum products, nitrate or Chromium) or naturally occurring (e.g., arsenic, salinity). Besides needing to understand where the groundwater is flowing, based on the other hydrologic properties, there are additional aquifer properties which affect how dissolved contaminants move with groundwater.

Hydrodynamic Dispersion

Hydrodynamic dispersivity (α_L, α_T) is an empirical factor which quantifies how much contaminants stray away from the path of the groundwater which is carrying it. Some of the contaminants will be "behind" or "ahead" the mean groundwater, giving rise to a longitudinal dispersivity (α_L), and some will be "to the sides of" the pure advective groundwater flow, leading to a transverse dispersivity (α_T). Dispersion in groundwater is due to the fact that each water "particle", passing beyond a soil particle, must choose where to go, whether left or right or up or down, so that the water "particles" (and their solute) are gradually spread in all directions around the mean path.

This is the "microscopic" mechanism, on the scale of soil particles. More important, on long distances, can be the macroscopic inhomogeneities of the aquifer, which can have regions of larger or smaller permeability, so that some water can find a preferential path in one direction, some other in a different direction, so that the contaminant can be spread in a completely irregular way, like in a (three-dimensional) delta of a river.

Dispersivity is actually a factor which represents our *lack of information* about the system we are simulating. There are many small details about the aquifer which are being averaged when using a macroscopic approach (e.g., tiny beds of gravel and clay in sand aquifers), they manifest themselves as an *apparent* dispersivity. Because of this, α is often claimed to be dependent on the length scale of the problem — the dispersivity found for transport through 1 m^3 of aquifer is different than that for transport through 1 cm^3 of the same aquifer material.

Molecular Diffusion

Diffusion is a fundamental physical phenomenon, by which Einstein explained Brownian motion, which describes the random thermal movement of molecules and small particles in gases and liquids. It is an important phenomenon for small distances (it is essential for the achievement of thermodinamic equilibria), but, as the time necessary to cover a distance by diffusion is proportional to the square of the distance itself, it is ineffective for spreading a solute over macroscopic distances. The diffusion coefficient, D, is typically quite small, and its effect can often be considered negligible (unless groundwater flow velocities are extremely low, as they are in clay aquitards).

It is important not to confuse diffusion with dispersion, as the former is a physical phenomenon and the latter is an empirical factor which is cast into a similar form as diffusion, because we already know how to solve that problem.

Retardation by Adsorption

The retardation factor is another very important feature that make the motion of the contaminant to deviate from the average groundwater motion. It is analogous to the retardation factor of chromatography. Unlike diffusion and dispersion, which simply spread the contaminant, the retardation factor changes its *global average velocity*, so that it can be much slower than that of water. This is due to a chemico-physical effect: the adsorption to the soil, which holds the contaminant back and does not allow it to progress until the quantity corresponding to the chemical adsorption equilibrium has been adsorbed. This effect is particularly important for less soluble contaminants, which thus can move even hundreds or thousands times slower than water. The effect of this phenomenon is that only more soluble species can cover long distances. The retardation factor depends on the chemical nature of both the contaminant and the aquifer.

GOVERNING EQUATIONS

Darcy's Law

Darcy's law is a Constitutive equation (empirically derived by Henri Darcy, in 1856) that states the amount of groundwater discharging through a given portion of aquifer is proportional to the cross-sectional area of flow, the hydraulic head gradient, and the hydraulic conductivity.

Groundwater Flow Equation

The groundwater flow equation, in its most general form, describes the movement of groundwater in a porous medium (aquifers and aquitards). It is known in mathematics as the diffusion equation, and has many analogs in other fields. Many solutions for groundwater flow problems were borrowed or adapted from existing heat transfer solutions. It is often derived from a physical basis using Darcy's law and a conservation of mass for a small control volume. The equation is often used to predict flow to wells, which have radial symmetry, so the flow equation is commonly solved in polar or cylindrical coordinates.

The Theis equation is one of the most commonly used and fundamental solutions to the groundwater flow equation; it can be used to predict the transient evolution of head due to the effects of pumping one or a number of pumping wells. The Thiem equation is a solution to the steady state groundwater flow equation (Laplace's Equation) for flow to a well. Unless there are large sources of water nearby (a river or lake), true steady-state is rarely achieved in reality.

Both above equations are used in aquifer tests (pump tests). The Hooghoudt equation is a groundwater flow equation applied to subsurface drainage by pipes, tile drains or ditches. An alternative subsurface drainage method is drainage by wells for which groundwater flow equations are also available.

CALCULATION OF GROUNDWATER FLOW

To use the groundwater flow equation to estimate the distribution of hydraulic heads, or the direction and rate of groundwater flow, this partial differential equation (PDE) must be solved. The most common means of analytically solving the diffusion equation in the hydrogeology literature are:

- Laplace, Hankel and Fourier transforms (to reduce the number of dimensions of the PDE),
- similarity transform (also called the Boltzmann transform) is commonly how the Theis solution is derived,

- separation of variables, which is more useful for non-Cartesian coordinates, and
- Green's functions, which is another common method for deriving the Theis solution — from the fundamental solution to the diffusion equation in free space.

No matter which method we use to solve the groundwater flow equation, we need both initial conditions (heads at time $(t) = 0$) and boundary conditions (representing either the physical boundaries of the domain, or an approximation of the domain beyond that point). Often the initial conditions are supplied to a transient simulation, by a corresponding steady-state simulation (where the time derivative in the groundwater flow equation is set equal to 0).

There are two broad categories of how the (PDE) would be solved; either analytical methods, numerical methods, or something possibly in between. Typically, analytic methods solve the groundwater flow equation under a simplified set of conditions *exactly*, while numerical methods solve it under more general conditions to an *approximation*.

Analytic Methods

Analytic methods typically use the structure of mathematics to arrive at a simple, elegant solution, but the required derivation for all but the simplest domain geometries can be quite complex (involving non-standard coordinates, conformal mapping, etc.). Analytic solutions typically are also simply an equation that can give a quick answer based on a few basic parameters. The Theis equation is a very simple (yet still very useful) analytic solution to the groundwater flow equation, typically used to analyze the results of an aquifer test or slug test.

Numerical Methods

The topic of numerical methods is quite large, obviously being of use to most fields of engineering and science in general. Numerical methods have been around much longer than computers have (In the 1920s Richardson developed some of the finite difference schemes still in use today, but they were calculated by hand, using paper and pencil, by human "calculators"), but they have become very important through the availability of fast and cheap personal computers.

There are two broad categories of numerical methods: gridded or discretized methods and non-gridded or mesh-free methods. In the common finite difference method and finite element method (FEM) the domain is completely gridded ("cut" into a grid or mesh of small elements). The analytic element method (AEM) and the boundary integral equation method (BIEM — sometimes also called BEM, or Boundary Element Method) are only discretized at boundaries or along flow elements (line sinks, area sources, etc.), the majority of the domain is mesh-free.

General Properties of Gridded Methods

Gridded Methods like finite difference and finite element methods solve the groundwater flow equation by breaking the problem area (domain) into many small elements (squares, rectangles, triangles, blocks, tetrahedra, etc.) and solving the flow equation for each element (all material properties are assumed constant or possibly linearly variable within an element), then linking together all the elements using conservation of mass across the boundaries between the elements (similar to the divergence theorem). This results in a system which overall approximates the groundwater flow equation, but exactly matches the boundary conditions (the head or flux is specified in the elements which intersect the boundaries).

Application of Finite Difference Models

MODFLOW is a well-known example of a general finite difference groundwater flow model. It is developed by the US Geological Survey as a modular and extensible simulation tool for modeling groundwater flow. It is free software developed, documented and distributed by the USGS. Many commercial products have grown up around it, providing graphical user interfaces to its input file based interface, and typically incorporating pre-and post-processing of user data. Many other models have been developed to work with MODFLOW input and output, making linked models which simulate several hydrologic processes possible (flow and transport models, surface water and groundwater models and chemical reaction models), because of the simple, well documented nature of MODFLOW.

Application of Finite Element Models

Finite Element programs are more flexible in design (triangular elements vs. the block elements most finite difference models use) and there are some programs available (SUTRA, a 2D or 3D density-dependent flow model by the USGS; Hydrus, a commercial unsaturated flow model; FEFLOW, a commercial modeling environment for subsurface flow, solute and heat transport processes; and COMSOL Multiphysics (FEMLAB) a commercial general modeling environment), but unless they are gaining in importance they are still not as popular in with practicing hydrogeologists as MODFLOW is.

Finite element models are more popular in university and laboratory environments, where specialized models solve non-standard forms of the flow equation (unsaturated flow, density dependent flow, coupled heat and groundwater flow, etc.)

Application of Finite Volume Models

The finite volume method is a method for representing and evaluating partial differential equations as algebraic equations. Similar to the finite difference method, values are calculated at discrete places on a meshed geometry.

"Finite volume" refers to the small volume surrounding each node point on a mesh. In the finite volume method, volume integrals in a partial differential equation that contain a divergence term are converted to surface integrals, using the divergence theorem. These terms are then evaluated as fluxes at the surfaces of each finite volume. Because the flux entering a given volume is identical to that leaving the adjacent volume, these methods are conservative. Another advantage of the finite volume method is that it is easily formulated to allow for unstructured meshes. The method is used in many computational fluid dynamics packages.

The FEHM software package is available free from Los Alamos National Laboratory and can be accessed at the FEHM Website. This versatile porous flow simulator includes capabilities to model multiphase, thermal, stress, and multicomponent reactive chemistry. Current work using this code includes simulation of methane hydrate formation, CO_2 sequestration, oil shale extraction, migration of both nuclear and chemical contaminants, environmental isotope migration in the unsaturated zone, and karst formation.

Other Methods

These include mesh-free methods like the Analytic Element Method (AEM) and the Boundary Element Method (BEM), which are closer to analytic solutions, but they do approximate the groundwater flow equation in some way. The BEM and AEM exactly solve the groundwater flow equation (perfect mass balance), while approximating the boundary conditions. These methods are more exact and can be much more elegant solutions (like analytic methods are), but have not seen as widespread use outside academic and research groups yet.

OCEANOGRAPHY

Oceanography also called oceanology or marine science, is the branch of Earth science that studies the ocean. It covers a wide range of topics, including marine organisms and ecosystem dynamics; ocean currents, waves, and geophysical fluid dynamics; plate tectonics and the geology of the sea floor; and fluxes of various chemical substances and physical properties within the ocean and across its boundaries. These diverse topics reflect multiple disciplines that oceanographers blend to further knowledge of the world ocean and understanding of processes within it: biology, chemistry, geology, meteorology, and physics as well as geography.

History

Humans first acquired knowledge of the waves and currents of the seas and oceans in pre-historic times. Observations on tides are recorded by Aristotle and Strabo. Early modern exploration of the oceans was primarily for cartography and mainly limited to its surfaces and of the creatures that fishermen brought up in nets, though depth soundings by lead line were taken.

Although Juan Ponce de León in 1513 first identified the Gulf Stream, and the current was well-known to mariners, Benjamin Franklin made the first scientific study of it and gave it its name. Franklin measured water temperatures during several Atlantic crossings and correctly explained the Gulf Stream's cause. Franklin and Timothy Folger printed the first map of the Gulf Stream in 1769-1770. When Louis Antoine de Bougainville, who voyaged between 1766 and 1769, and James Cook, who voyaged from 1768 to 1779, carried out their explorations in the South Pacific, information on the oceans themselves formed part of the reports. James Rennell wrote the first scientific textbooks about currents in the Atlantic and Indian oceans during the late 18th and at the beginning of 19th century. Sir James Clark Ross took the first modern sounding in deep sea in 1840, and Charles Darwin published a paper on reefs and the formation of atolls as a result of the second voyage of HMS *Beagle* in 1831-6. Robert FitzRoy published a report in four volumes of the three voyages of the *Beagle*. In 1841–1842 Edward Forbes undertook dredging in the Aegean Sea that founded marine ecology.

As first superintendent of the United States Naval Observatory (1842–1861) Matthew Fontaine Maury devoted his time to the study of marine meteorology, navigation, and charting prevailing winds and currents. His *Physical Geography of the Sea*, 1855 was the first textbook of oceanography. Many nations sent oceanographic observations to Maury at the Naval Observatory, where he and his colleagues evaluated the information and gave the results worldwide distribution.

The steep slope beyond the continental shelves was discovered in 1849. The first successful laying of transatlantic telegraph cable in August 1858 confirmed the presence of an underwater "telegraphic plateau" mid-ocean ridge. After the middle of the 19th century, scientific societies were processing a flood of new terrestrial botanical and zoological information.

In 1871, under the recommendations of the Royal Society of London, the British government sponsored an expedition to explore world's oceans and conduct scientific investigations. Under that sponsorship the Scots Charles Wyville Thompson and Sir John Murray launched the Challenger expedition (1872–1876). The results of this were published in 50 volumes covering biological, physical and geological aspects. 4417 new species were discovered.

Other European and American nations also sent out scientific expeditions (as did private individuals and institutions). The first purpose built oceanographic ship, the "Albatros" was built in 1882. The four-month 1910 North Atlantic expedition headed by Sir John Murray and Johan Hjort was at that time the most ambitious research oceanographic and marine zoological project ever, and led to the classic 1912 book *The Depths of the Ocean*. Oceanographic institutes dedicated to the study of oceanography were founded. In the United States, these included the Scripps Institution of Oceanography in 1892, Woods Hole Oceanographic Institution in 1930, Virginia Institute of Marine Science in 1938, Lamont-Doherty Earth Observatory at Columbia University, and the School of Oceanography at University of Washington. In Britain, there is a major research institution: National Oceanography Centre, Southampton which is the successor to the Institute of Oceanography. In Australia, CSIRO Marine and Atmospheric Research, known as CMAR, is a leading centre. In 1921 the International Hydrographic Bureau (IHB) was formed in Monaco.

In 1893, Fridtjof Nansen allowed his ship "Fram" to be frozen in the Arctic ice. As a result he was able to obtain oceanographic data as well as meteorological and astronomical data. The first international organization of oceanography was created in 1902 as the International Council for the Exploration of the Sea. The first acoustic measurement of sea depth was made in 1914. Between 1925 and 1927 the "Meteor" expedition gathered 70,000 ocean depth measurements using an echo sounder, surveying the Mid atlantic ridge. The Great Global Rift, running along the Mid Atlantic Ridge, was discovered by Maurice Ewing and Bruce Heezen in 1953 while the mountain range under the Arctic was found in 1954 by the Arctic Institute of the USSR. The theory of seafloor spreading was developed in 1960 by Harry Hammond Hess. The Ocean Drilling Project started in 1966. Deep sea vents were discovered in 1977 by John Corlis and Robert Ballard in the submersible "Alvin".

In the 1950s, Auguste Piccard invented the bathyscaphe and used the "Trieste" to investigate the ocean's depths. The nuclear submarine Nautilus made the first journey under the ice to the North Pole in 1958. In 1962 there was the first deployment of FLIP (Floating Instrument Platform), a 355 foot spar buoy.

Then, in 1966, the U.S. Congress created a *National Council for Marine Resources and Engineering Development*. NOAA was put in charge of exploring and studying all aspects of Oceanography in the USA. It also enabled the National Science Foundation to award *Sea Grant College* funding to multidisciplinary researchers in the field of oceanography. From the 1970s, there has been much emphasis on the application of large scale computers to oceanography to allow numerical predictions of ocean conditions and as a part of overall environmental change prediction. An oceanographic buoy array was established in the Pacific to allow prediction of El Niño events.

1990 saw the start of the World Ocean Circulation Experiment (WOCE) which continued until 2002. Geosat seafloor mapping data became available in 1995. In 1942, Sverdrup and Fleming published "The Ocean" which was a major landmark. "The Sea" (in three volumes covering physical oceanography, seawater and geology) edited by M.N. Hill was published in 1962 while the "Encyclopedia of Oceanography" by Rhodes Fairbridge was published in 1966.

Connection to the Atmosphere

The study of the oceans is linked to understanding global climate changes, potential global warming and related biosphere concerns. The atmosphere and ocean are linked because of evaporation and precipitation as well as thermal flux (and solar insolation). Wind stress is a major driver of ocean currents while the ocean is a sink for atmospheric carbon dioxide. Our planet is invested with two great oceans; one visible, the other invisible; one underfoot, the other overhead; one entirely envelopes it, the other covers about two thirds of its surface —Matthew F. Maury, *The Physical Geography of the Seas and Its Meteorology (1855)*

Branches

The study of oceanography is divided into branches:

- Biological oceanography, or marine biology, is the study of the plants, animals and microbes of the oceans and their ecological interaction with the ocean;
- Chemical oceanography, or marine chemistry, is the study of the chemistry of the ocean and its chemical interaction with the atmosphere;

- Geological oceanography, or marine geology, is the study of the geology of the ocean floor including plate tectonics;
- Physical oceanography, or marine physics, studies the ocean's physical attributes including temperature-salinity structure, mixing, waves, internal waves, surface tides, internal tides, and currents. Of particular interest is the behavior of sound (acoustical oceanography), light (optical oceanography) and radio waves in the ocean.

These branches reflect the fact that many oceanographers are first trained in the exact sciences or mathematics and then focus on applying their interdisciplinary knowledge, skills and abilities to oceanography. Data derived from the work of Oceanographers is used in marine engineering, in the design and building of oil platforms, ships, harbours, and other structures that allow us to use the ocean safely. Oceanographic data management is the discipline ensuring that oceanographic data both past and present are available to researchers.

CHEMICAL OCEANOGRAPHY

Chemical oceanography is the study of ocean chemistry: the behavior of the chemical elements within the Earth's oceans. The ocean is unique in that it contains-in greater or lesser quantities-nearly every element in the periodic table.

Much of chemical oceanography describes the cycling of these elements both within the ocean and with the other spheres of the Earth system. These cycles are usually characterised as quantitative fluxes between constituent reservoirs defined within the ocean system and as residence times within the ocean. Of particular global and climatic significance are the cycles of the biologically active elements such as carbon, nitrogen, and phosphorus as well as those of some important trace elements such as iron.

Another important area of study in chemical oceanography is the behaviour of isotopes and how they can be used as tracers of past and present oceanographic and climatic processes. For example, the incidence of ^{18}O (the heavy isotope of oxygen) can be used as an indicator of polar ice sheet extent, and boron isotopes are key indicators of the pH and CO_2 content of oceans in the geologic past.

MARINE BIOLOGY

Marine biology is the scientific study of organisms in the ocean or other marine or brackish bodies of water. Given that in biology many phyla, families and genera have some species that live in the sea and others that live on land, marine biology classifies species based on the environment rather than on taxonomy. Marine biology differs from marine ecology as marine ecology is focused on how organisms interact with each other and the environment, and biology is the study of the organisms themselves. Marine life is a vast resource, providing food, medicine, and raw materials, in addition to helping to support recreation and tourism all over the world. At a fundamental level, marine life helps determine the very nature of our planet. Marine organisms contribute significantly to the oxygen cycle, and are involved in the regulation of the Earth's climate. Shorelines are in part shaped and protected by marine life, and some marine organisms even help create new land.

Marine biology covers a great deal, from the microscopic, including most zooplankton and phytoplankton to the huge cetaceans (whales) which reach up to a reported 48 meters (125 feet) in length. The habitats studied by marine biology include everything from the tiny layers of surface water in which organisms and abiotic items may be trapped in surface tension between the ocean and atmosphere, to the depths of the oceanic trenches, sometimes 10,000 meters or more beneath the surface of the ocean. It studies habitats such as coral reefs, kelp forests, tidepools, muddy, sandy and rocky bottoms, and the open ocean (pelagic) zone, where solid objects are rare and the surface of the water is the only visible boundary.

A large amount of all life on Earth exists in the oceans. Exactly how large the proportion is unknown, since many ocean species are still to be discovered. While the oceans comprise about 71% of the Earth's surface, due to their depth they encompass about 300 times the habitable volume of the terrestrial habitats on Earth.

Many species are economically important to humans, including food fish. It is also becoming understood that the well-being of marine organisms and other organisms are linked in very fundamental ways. The human body of knowledge regarding the relationship between life in the sea and important cycles is rapidly growing, with new discoveries being made nearly every day. These cycles include those of matter (such as the carbon cycle) and of air (such as Earth's respiration, and movement of energy through ecosystems including the ocean). Large areas beneath the ocean surface still remain effectively unexplored.

Subfields

The marine ecosystem is large, and thus there are many subfields of marine biology. Most involve studying specializations of particular animal groups, such as phycology, invertebrate zoology and ichthyology. Other subfields study the physical effects of continual immersion in sea water and the ocean in general, adaptation to a salty environment, and the effects of changing various oceanic properties on marine life. A subfield of marine biology studies the relationships between oceans and ocean life, and global warming and environmental issues (such as carbon dioxide displacement).

Recent marine biotechnology has focused largely on marine biomolecules, especially proteins, that may have uses in medicine or engineering. Marine environments are the home to many exotic biological materials that may inspire biomimetic materials.

Related Fields

Marine biology is a branch of oceanography and is closely linked to biology. It also encompasses many ideas from ecology. Fisheries science and marine conservation can be considered partial offshoots of marine biology (as well as environmental studies).

Lifeforms

Microscopic Life

Microscopic life undersea is incredibly diverse and still poorly understood. For example, the role of viruses in marine ecosystems is barely being explored even in the beginning of the 21st century. The role of phytoplankton is better understood due to their critical position as the most numerous primary producers on Earth. Phytoplankton are categorized into cyanobacteria (also called blue-green algae/bacteria), various types of algae (red, green, brown, and yellow-green), diatoms, dinoflagellates, euglenoids, coccolithophorids, cryptomonads, chrysophytes, chlorophytes, prasinophytes, and silicoflagellates.

Zooplankton tend to be somewhat larger, and not all are microscopic. Many Protozoa are zooplankton, including dinoflagellates, zooflagellates, foraminiferans, and radiolarians. Some of these (such as dinoflagellates) are also phytoplankton; the distinction between plants and animals often breaks down in very small organisms. Other zooplankton include cnidarians, ctenophores, chaetognatbs, molluscs, arthropods, urochordates, and annelids such as polychaetes. Many larger animals begin their life as zooplankton before they become large enough to take their familiar forms. Two examples are fish larvae and sea stars (also called starfish).

Plants and Algae

Plant life is widespread and very diverse under the ocean. Microscopic photosynthetic algae contribute a larger proportion of the worlds photosynthetic output than all the terrestrial forests combined. Most of the niche occupied by sub plants on land is actually occupied by macroscopic algae in the ocean, such as *Sargassum* and kelp, which are commonly known as seaweeds that creates kelp forests. The non algae plants that survive in the sea are often found in shallow waters, such as the seagrasses (examples of which are eelgrass, *Zostera*, and turtle grass, *Thalassia*). These plants have adapted to the high salinity of the ocean environment. The intertidal zone is also a good place to find plant life in the sea, where mangroves or cordgrass or beach grass might grow. Microscopic algae and plants provide important habitats for life, sometimes acting as hiding and foraging places for larval forms of larger fish and invertebrates.

Marine Invertebrates

As on land, invertebrates make up a huge portion of all life in the sea. Invertebrate sea life includes Cnidaria such as jellyfish and sea anemones; Ctenophora; sea worms including the phyla Platyhelminthes, Nemertea, Annelida, Sipuncula, Echiura, Chaetognatha, and Phoronida; Mollusca including shellfish, squid, octopus; Arthropoda including Chelicerata and

Crustacea; Porifera; Bryozoa; Echinodermata including starfish; and Urochordata including sea squirts or tunicates.

Fish

Fish have evolved very different biological functions from other large organisms. Fish anatomy includes a two-chambered heart, operculum, swim bladder, scales, fins, lips, eyes and secretory cells that produce mucous. Fish breathe by extracting oxygen from water through their gills. Fins propel and stabilize the fish in the water. Well known fish include: sardines, anchovy, ling cod, clownfish (also known as anemonefish), and bottom fish which include halibut or ling cod. Predators include sharks and barracuda.

Reptiles

Reptiles which inhabit or frequent the sea include sea turtles, sea snakes, terrapins, the marine iguana, and the saltwater crocodile. Most extant marine reptiles, except for some sea snakes, are oviparous and need to return to land to lay their eggs. Thus most species, excepting sea turtles, spend most of their lives on or near land rather than in the ocean. Despite their marine adaptations, most sea snakes prefer shallow waters not far from land, around islands, especially waters that are somewhat sheltered, as well as near estuaries. Some extinct marine reptiles, such as ichthyosaurs, evolved to be viviparous and had no requirement to return to land.

Seabirds

Seabirds are species of birds adapted to living in the marine environment, examples including albatross, penguins, gannets, and auks. Although they spend most of their lives in the ocean, species such as gulls can often be found thousands of miles inland.

Marine Mammals

There are five main types of marine mammals.

- Cetaceans include toothed whales (Suborder Odontoceti), such as the Sperm Whale, dolphins, and porpoises such as the Dall's porpoise. Cetaceans also include baleen whales (Suborder Mysticeti), such as the Gray Whale, Humpback Whale, and Blue Whale.

- Sirenians include manatees, the Dugong, and the extinct Steller's Sea Cow.
- Seals (Family Phocidae), sea lions (Family Otariidae-which also include the fur seals), and the Walrus (Family Odobenidae) are all considered pinnipeds.
- The Sea Otter is a member of the Family Mustelidae, which includes weasels and badgers.
- The Polar Bear (Family Ursidae) is sometimes considered a marine mammal because of its dependence on the sea.

OCEANIC HABITATS

Reefs

Reefs comprise some of the densest and most diverse habitats in the world. The best-known types of reefs are tropical coral reefs which exist in most tropical waters; however, reefs can also exist in cold water. Reefs are built up by corals and other calcium-depositing animals, usually on top of a rocky outcrop on the ocean floor. Reefs can also grow on other surfaces, which has made it possible to create artificial reefs. Coral reefs also support a huge community of life, including the corals themselves, their symbiotic zooxanthellae, tropical fish and many other organisms. Much attention in marine biology is focused on coral reefs and the El Niño weather phenomenon. In 1998, coral reefs experienced the most severe mass bleaching events on record, when vast expanses of reefs across the world died because sea surface temperatures rose well above normal. Some reefs are recovering, but scientists say that between 50% and 70% of the world's coral reefs are now endangered and predict that global warming could exacerbate this trend.

Deep Sea and Trenches

The deepest recorded oceanic trenches measure to date is the Mariana Trench, near the Philippines, in the Pacific Ocean at 10,924 m (35,838 ft). At such depths, water pressure is extreme and there is no sunlight, but some life still exists. A white flatfish, a shrimp and a jellyfish were seen by the American crew of the bathyscaphe *Trieste* when it dove to the bottom in 1960. Other notable oceanic trenches include Monterey Canyon, in the eastern Pacific, the Tonga Trench in the southwest at

10,882 m (35,702 ft), the Philippine Trench, the Puerto Rico Trench at 8,605 m (28,232 ft), the Romanche Trench at 7,760 m (24,450 ft), Fram Basin in the Arctic Ocean at 4,665 m (15,305 ft), the Java Trench at 7450 m (24,442 ft), and the South Sandwich Trench at 7,235 m (23,737 ft).

In general, the deep sea is considered to start at the aphotic zone, the point where sunlight loses its power of transference through the water. Many life forms that live at these depths have the ability to create their own light a unique evolution known as bio-luminescence. Marine life also flourishes around seamounts that rise from the depths, where fish and other sea life congregate to spawn and feed. Hydrothermal vents along the mid-ocean ridge spreading centers act as oases, as do their opposites, cold seeps. Such places support unique biomes and many new microbes and other lifeforms have been discovered at these locations.

Open Ocean

The open ocean is relatively unproductive because of a lack of nutrients, yet because it is so vast, in total it produces the most primary productivity. Much of the aphotic zone's energy is supplied by the open ocean in the form of detritus. The open ocean consists mostly of jellyfish and its predators such as the mola mola.

Intertidal and Shore

Intertidal zones, those areas close to shore, are constantly being exposed and covered by the ocean's tides. A huge array of life lives within this zone. Shore habitats span from the upper intertidal zones to the area where land vegetation takes prominence. It can be underwater anywhere from daily to very infrequently. Many species here are scavengers, living off of sea life that is washed up on the shore. Many land animals also make much use of the shore and intertidal habitats. A subgroup of organisms in this habitat bores and grinds exposed rock through the process of bioerosion.

Distribution Factors

An active research topic in marine biology is to discover and map the life cycles of various species and where they spend their time. Marine biologists study how the ocean currents, tides and many other oceanic factors affect ocean lifeforms, including their growth, distribution and well-being. This has only recently become technically feasible with advances in GPS and newer underwater visual devices.

Most ocean life breeds in specific places, nests or not in others, spends time as juveniles in still others, and in maturity in yet others. Scientists know little about where many species spend different parts of their life cycles. For example, it is still largely unknown where sea turtles and some sharks travel. Tracking devices do not work for some life forms, and the ocean is not friendly to technology. This is important to scientists and fishermen because they are discovering that by restricting commercial fishing in one small area they can have a large impact in maintaining a healthy fish population in a much larger area far away.

MARINE GEOLOGY

Marine geology involves geophysical, geochemical, sedimentological and paleontological investigations of the ocean floor and coastal margins. Marine geology has strong ties to physical oceanography and plate tectonics. Marine geological studies were of extreme importance in providing the critical evidence for sea floor spreading and plate tectonics in the years following World War II. The deep ocean floor is the last essentially unexplored frontier and detailed mapping in support of both military (submarine) objectives and economic (petroleum and metal mining) objectives drives the research.

Overview

The Ring of Fire around the Pacific Ocean with its attendant intense volcanism and seismic activity poses a major threat for disastrous earthquakes, tsunamis and volcanic eruptions. Any *early warning* systems for these disastrous events will require a more detailed understanding of marine geology of coastal and island arc environments. The study of littoral and deep sea sedimentation and the precipitation and dissolution rates of calcium carbonate in various marine environments has important implications for global climate change.

The discovery and continued study of mid-ocean rift zone volcanism and hydrothermal vents, first in the Red Sea and later along the East Pacific Rise and the Mid-Atlantic Ridge systems were and continue to be important areas of marine geological research. The extremophile organisms discovered living within and adjacent to those hydrothermal systems have had a pronounced impact on our understanding of life on Earth and potentially the origin of life within such an environment.

Oceanic trenches are hemispheric-scale long but narrow topographic depressions of the sea floor. They also are the deepest parts of the ocean floor. The Mariana Trench (or Marianas Trench) is the deepest known submarine trench, and the deepest location in the Earth's crust itself. A subduction zone where the Pacific Plate is being subducted under the Philippine Sea Plate. The bottom of the trench is further below sea level than Mount Everest is above sea level.

6

The Pedosphere

The pedosphere is the outermost layer of the Earth that is composed of soil and subject to soil formation processes. It exists at the interface of the lithosphere, atmosphere, hydrosphere and biosphere.

The pedosphere acts as the mediator of chemical and biogeochemical flux into and out of these respective systems and is made up of gaseous, mineralic, fluid and biologic components. The pedosphere lies within the Critical Zone, a broader interface that includes vegetation, pedosphere, groundwater aquifer systems, regolith and finally ends at some depth in the bedrock where the biosphere and hydrosphere cease to make significant changes to the chemistry at depth. As part of the larger global system, any particular environment in which soil forms is influenced solely by its geographic position on the globe as climatic, geologic, biologic and anthropogenic changes occur with changes in longitude and latitude.

The pedosphere lies below the vegetative cover of the biosphere and above the hydrosphere and lithosphere. The soil forming process (pedogenesis) can begin without the aid of biology but is significantly quickened in the presence of biologic reactions. Soil formation begins with the chemical and/or physical breakdown of minerals to form the initial material that overlies the bedrock substrate. Biology quickens this by secreting acidic compounds (dominantly fulvic acids) that help break rock apart. Particular biologic pioneers are lichen, mosses and seed bearing plants but many other inorganic reactions take place that diversify the chemical makeup of the early soil layer. Once weathering and decomposition products accumulate, a coherent soil body allows the migration of fluids both vertically and laterally through the soil profile causing ion exchange between solid, fluid and gaseous phases. As time progresses, the bulk geochemistry of the soil layer will deviate away from the initial composition of the bedrock and will evolve to a chemistry that reflects the type of reactions that take place in the soil.

Lithosphere

The primary conditions for soil development are controlled by the chemical composition of the rock that the soil will eventually be forming on. Rock types that form the base of the soil profile are often either sedimentary (carbonate or siliceous), igneous or metaigneous (metamorphosed igneous rocks) or volcanic and metavolcanic rocks. The rock type and the processes that lead to its exposure at the surface are controlled by the regional geologic setting of the specific area under study, which revolve around the underlying theory of plate tectonics, subsequent deformation, uplift, subsidence and deposition.

Metaigneous and metavolcanic rocks form the largest component of cratons and are high in silica. Igneous and volcanic rocks are also high in silica but with non-metamorphosed rock, weathering becomes faster and the mobilization of ions is more widespread.

Rocks high in silica produce silicic acid as a weathering product. There are few rock types that lead to localized enrichment of some of the biologically limiting elements like phosphorus (P) and nitrogen (N). Phosphatic shale (<15% P_2O_5) and phosphorite (>15% P_2O_5) form in anoxic deep water basins that preserve organic material. Greenstone (metabasalt), phyllite and schist release up to 30-50% of the nitrogen pool. Thick successions of carbonate rocks are often deposited on craton margins during sea level rise. The widespread dissolution of carbonate and evaporate minerals leads to elevated levels of Mg^{2+}, HCO_3^-, Sr^{2+}, Na^+, Cl^- and SO_4^{2-} ions in aqueous solution.

Weathering and Dissolution of Minerals

The process of soil formation is dominated by chemical weathering of silicate minerals, aided by acidic products of pioneering plants and organisms as well as carbonic acid inputs from the atmosphere. Carbonic acid is produced in the atmosphere and soil layers through the carbonation reaction.

$H_2O + CO_2 \rightarrow H^+ + HCO_3^- \rightarrow H_2CO_3$

This is the dominant form of chemical weathering and aides in the breakdown of carbonate minerals like calcite and dolomite and silicate minerals like feldspar. The breakdown of the Na-feldspar, albite, by carbonic acid to form kaolinite clay is as follows:

$2NaAlSi_3O_8 + 2H_2CO_3 + 9H_2O \rightarrow 2Na^+ + 2HCO_3^- + 4H_4SiO_4 + Al_2Si_2O_5(OH)_4$

Evidence of this reaction in the field would be elevated levels of bicarbonate (HCO_3^-), sodium and silica ions in the water runoff. The breakdown of carbonate minerals:

$CaCO_3 + H_2CO_3 \rightarrow Ca^{2+} + 2HCO_3^-$ or $CaCO_3 \rightarrow Ca^{2+} + CO_3^{2-}$

The further dissolution of carbonic acid (H_2CO_3) and bicarbonate (HCO_3) produces CO_2 gas. Oxidization is also a major contributor to the breakdown of many silicate minerals and formation of secondary minerals (diagenesis) in the early soil profile. Oxidation of Olivine ($FeMgSiO_2$) releases Fe, Mg and Si ions. The Mg is soluble in water and is carried in the runoff but the Fe often reacts with oxygen to precipitate Fe_2O_3 (hematite), the oxidized state of iron oxide. Sulfur, a byproduct of decaying organic material will also react to Fe to form pyrite (FeS_2) but often in reducing environments. Pyrite dissolution leads to high pH levels due to elevated H+ ions and further precipitation of Fe_2O_3 ultimately changing the redox conditions of the environment.

Biosphere

Inputs from the biosphere may begin with lichen and other microorganisms that secrete oxalic acid. These microorganisms, associated with the lichen community or independently inhabiting rocks, include a number of blue-green algae, green algae, various fungi, and numerous bacteria. Lichen has long been viewed as the pioneers of soil development as the following statement suggests:

"The initial conversion of rock into soil is carried on by the pioneer lichens and their successors, the mosses, in which the hair-like rhizoids assume the role of roots in breaking down the surface into fine dust "

However, lichens are not necessarily the only pioneering organisms nor the earliest form of soil formation as it has been documented that seed-bearing plants may occupy an area and colonize quicker than lichen. Also, eolian sedimentation can produce high rates of sediment accumulation. Nonetheless, lichen can certainly withstand harsher conditions than most vascular plants and although they have slower colonization rates, do form the dominant group in alpine regions.

Acids released from plant roots include acetic and citric acids. During the decay of organic matter Phenolic acids are released from plant matter and humic and fulvic acids are released by soil microbes. These organic acids speed up chemical weathering by combining with some of the weathering products in a process known as chelation. In the soil profile, the organic acids are often concentrated at the top while carbonic acid plays a larger role towards the bottom or below in the aquifer.

As the soil column develops further into thicker accumulations, larger animals come to inhabit the soil and continue to alter the chemical evolution of their respective niche. Earthworms aerate the soil and convert large amounts of organic matter into rich humus, improving soil fertility. Small burrowing mammals store food, grow young and may hibernate in the pedosphere altering the course of soil evolution. Large mammalian herbivores above ground transport nutrients in form of nitrogen-rich waste and phosphorus-rich antlers while predators leave phosphorus-rich piles of bones on the soil surface, leading the localized enrichment of the soil below.

Redox Conditions in Wetland Soils

Nutrient cycling in lakes and freshwater wetlands depends heavily on redox conditions. Under a few millimeters of water heterotrophic bacteria metabolize and consume oxygen. They therefore deplete the soil of oxygen and create the need for anaerobic respiration. Some anaerobic microbial processes include denitrification, sulfate reduction and methanogenesis and are responsible for the release of N_2 (nitrogen), H_2S (hydrogen sulfide) and CH_4 (methane). Other anaerobic microbial processes are linked to changes in the oxidation state of iron and manganese. As a result of anaerobic decomposition, the soil stores large amounts of organic carbon because decomposition is incomplete.

The redox potential describes which way chemical reactions will proceed in oxygen deficient soils and controls the nutrient cycling in flooded systems. Redox potential, or reduction potential, is used to express the likelihood of an environment to receive electrons and therefore become reduced. For example, if a system already has plenty of electrons (anoxic, organic-rich shale) it is reduced and will likely donate electrons to a part of the system that has a low concentration of electrons, or an oxidized environment, to equilibrate to the chemical gradient. The oxidized environment has high redox potential, whereas the reduced environment has a low redox potential.

The redox potential is controlled by the oxidation state of the chemical species, pH and the amount of oxygen (O_2) there is in the system. The oxidizing environment accepts electrons because of the presence of O_2, which acts as electron acceptors:

$$O_2 + 4e^- + 4H^+ \rightarrow H_2O$$

This equation will tend to move to the right in acidic conditions which causes higher redox potentials to be found at lower pH levels. Bacteria, heterotrophic organisms, consume oxygen while decomposing organic material which depletes the soils of oxygen, thus increasing the redox potential. In low redox conditions the deposition of ferrous iron (Fe^{2+}) will increase with decreasing decomposition rates, thus preserving organic remains and depositing humus. At high redox potential, the oxidized form of iron, ferric iron (Fe^{3+}), will be deposited commonly as hematite. By using analytical geochemical tools such as x-ray fluorescence (XRF) or inductively coupled mass spectroscopy (ICP-MS) the two forms of Fe (Fe^{2+} and Fe^{3+}) can be measured in ancient rocks therefore determining the redox potential for ancient soils.

Such a study was done on Permian through Triassic rocks (300-200 million years old) in Japan and British Colombia. The geologists found hematite throughout the early and middle Permian but began to find the reduced form of iron in pyrite within the ancient soils near the end of the Permian and into the Triassic. This suggests that conditions became less oxygen rich, even anoxic, during the late Permian which eventually lead to the greatest extinction in earth's history, the P-T extinction.

Decomposition in anoxic or reduced soils is also carried out by sulfur-reducing bacteria which, instead of O_2 use SO_4^{2-}as an electron acceptor and produce hydrogen sulfide (H_2S) and carbon dioxide in the process:

$$2H^+ + SO_4^{2-} + 2(CH_2O) \rightarrow 2CO_2 + H_2S + 2H_2O$$

The H_2S gas percolates upwards and reacts with Fe^{2+} and precipitates pyrite, acting as a trap for the toxic H_2S gas. However, H_2S is still a large fraction of emissions from wetland soils. In most freshwater wetlands there is little sulfate (SO_4^{2-}) so methanogenesis becomes the dominant form of decomposition by methanogenic bacteria only when sulfate is depleted. Acetate, a compound that is a byproduct of fermenting cellulose is split by methanogenic bacteria to produce methane (CH_4) and carbon dioxide (CO_2), which are released to the atmosphere. Methane is also released during the reduction of CO_2 by the same bacteria.

Atmosphere

In the pedosphere it is safe to assume that gases are in equilibrium with the atmosphere. Because plant roots and soil microbes release CO_2 to the soil, the concentration of bicarbonate(HCO_3) in soil waters is much greater than that in equilibrium with the atmosphere, the high concentration of CO_2 and the occurrence of metals in soil solutions results in lower pH levels in the soil. Gases that escape from the pedosphere to the atmosphere include the gaseous byproducts of carbonate dissolution, decomposition, redox reactions and microbial photosynthesis.

The main inputs from the atmosphere are aeolian sedimentation, rainfall and gas diffusion. Eolian sedimentation includes anything that can be entrained by wind or that stays suspended, seemingly indefinitely, in air and includes a wide variety of aerosol particles, biological particles like pollen and dust to pure quartz sand. Nitrogen is the most abundant constituent in rain, as water vapour utilizes aerosol particles to nucleate rain droplets.

Soil in Forests

Soil is well developed in the forest as suggested by the thick humus layers, rich diversity of large trees and animals that live there. In forests, precipitation exceeds evapotranspiration which results in an excess of water that percolates downward through the soil layers. Slow rates of decomposition leads to large amounts of fulvic acid, greatly enhancing chemical weathering. The downward percolation, in conjunction with chemical weathering leaches magnesium (Mg), iron (Fe), and aluminum (Al) from the soil and transports them downward, a process known as podzolization. This process leads to marked contrasts in the appearance and chemistry of the soil layers.

Soil in the Tropics

Tropical forests (rainforests) receive more insolation and rainfall over longer growing seasons than any other environment on earth. With these elevated temperatures, insolation and rainfall, biomass is extremely productive leading to carbon production as much as $800gCm^{2}\cdot yr^{-}$. Higher temperatures and larger amounts of water contribute to higher rates of chemical weathering. Increased rates of decomposition cause smaller amounts of fulvic acid to percolate and leach metals from the zone of active weathering. Thus, in stark contrast to soil in forests, tropical forests have little to no podzolization and therefore do not have marked visual and chemical contrasts with the soil layers. Instead, the mobile metals Mg, Fe and Al are precipitated as oxide minerals giving the soil a rusty red colour.

Soil in Grasslands and Deserts

Precipitation in grasslands is equal to or less than evapotranspiration and causes soil development to operate in relative drought. Leaching and migration of weathering products is therefore decreased. Large amounts of evaporation causes buildup of calcium (Ca) and other large cations flocculate clay minerals and fulvic acids in the upper soil profile. Impermeable clay limits downward percolation of water and fulvic acids, reducing chemical weathering and podzolization. The depth to the maximum concentration of clay increases in areas of increased precipitation and leaching. When leaching is decreased, the Ca precipitates as calcite ($CaCO_3$) in the lower soil levels, a layer known as caliche.

Deserts behave similarly to grasslands but operate in constant drought as precipitation is less than evapotranspiration. Chemical weathering proceeds more slowly than in grasslands and beneath the caliche layer may be a layer of gypsum and halite. To study soils in deserts, pedologists have used the concept of chronosequences to relate timing and development of the soil layers. It has been shown that P is leached very quickly from the system and therefore decreases with increasing age. Furthermore, carbon buildup in the soils is decrease due to slower decomposition rates. As a result, the rates of carbon circulation in the biogeochemical cycle is decreased.

SOIL SCIENCE

Soil science is the study of soil as a natural resource on the surface of the earth including soil formation, classification and mapping; physical, chemical, biological, and fertility properties of soils; and these properties in relation to the use and management of soils.

Sometimes terms which refer to branches of soil science, such as pedology (formation, chemistry, morphology and classification of soil) and edaphology (influence of soil on organisms, especially plants), are used as if synonymous with soil science. The diversity of names associated with this discipline is related to the various associations concerned. Indeed, engineers, agronomists, chemists, geologists, geographers, ecologists, biologists, microbiologists, sylviculturists, sanitarians, archaeologists, and specialists in regional planning, all contribute to further knowledge of soils and the advancement of the soil sciences.

Fields of Study

Soil occupies the pedosphere, one of Earth's spheres that the geosciences use to organize the Earth conceptually. This is the conceptual perspective of pedology and edaphology, the two main branches of soil science. Pedology is the study of soil in its natural setting. Edaphology is the study of soil in relation to soil-dependent uses. Both branches apply a combination of soil physics, soil chemistry, and soil biology. Due to the numerous interactions between the biosphere, atmosphere and hydrosphere that are hosted within the pedosphere, more integrated, less soil-centric concepts are also valuable. Many concepts essential to understanding soil come from individuals not identifiable strictly as soil scientists. This highlights the interdisciplinary nature of soil concepts.

Research

Dependence on and curiosity about soil, exploring the diversity and dynamic of this resource continues to yield fresh discoveries and insights. New avenues of soil research are compelled by a need to understand soil in the context of climate change, greenhouse gases, and carbon sequestration. Interest in maintaining the planet's biodiversity and in exploring past cultures has also stimulated renewed interest in achieving a more refined understanding of soil.

Mapping

Most empirical knowledge of soil in nature comes from soil survey efforts. Soil survey, or soil mapping, is the process of determining the soil types or other properties of the soil cover over a landscape, and mapping them for others to understand and use. It relies heavily on distinguishing the individual influences of the five classic soil forming factors. This effort draws upon geomorphology, physical geography, and analysis of vegetation and land-use patterns. Primary data for the soil survey are acquired by field sampling and supported by remote sensing.

Classification

As of 2006, the World Reference Base for Soil Resources, via its Land & Water Development division, is the pre-eminent soil classification system. It replaces the previous FAO soil classification.

The WRB borrows from modern soil classification concepts, including USDA soil taxonomy. The classification is based mainly on soil morphology as an expression pedogenesis. A major difference with USDA soil taxonomy is that soil climate is not part of the system, except insofar as climate influences soil profile characteristics.

Many other classification schemes exist, including vernacular systems. The structure in vernacular systems are either nominal, giving unique names to soils or landscapes, or descriptive, naming soils by their characteristics such as red, hot, fat, or sandy.

Soils are distinguished by obvious characteristics, such as physical appearance (e.g., colour, texture, landscape position), performance (e.g., production capability, flooding), and accompanying vegetation. A vernacular distinction familiar to many is classifying texture as heavy or light. Light soil content and better structure, take less effort to turn and cultivate. Contrary to popular belief, light soils do not weigh less than heavy soils on an air dry basis nor do they have more porosity.

History

Vasily Dokuchaev, a Russian geologist, geographer and early soil scientist, is credited with identifying soil as a resource whose distinctness and complexity deserved to be separated conceptually from geology and crop production and treated as a whole.

Previously, soil had been considered a product of chemical transformations of rocks, a dead substrate from which plants derive nutritious elements. Soil and bedrock were in fact equated. Dokuchaev considers the soil as a natural body having its own genesis and its own history of development, a body with complex and multiform processes taking place within it. The soil is considered as different from bedrock. The latter becomes soil under the influence of a series of soil-formation factors (climate, vegetation, country, relief and age). According to him, soil should be called the "daily" or outward horizons of rocks regardless of the type; they are changed naturally by the common effect of water, air and various kinds of living and dead organisms.

A 1914 encyclopedic definition: "the different forms of earth on the surface of the rocks, formed by the breaking down or weathering of rocks". serves to illustrate the historic view of soil which persisted from the 19th century. Dokuchaev's late 19th century soil concept developed in the 20th century to one of soil as earthy material that has been altered by living processes. A corollary concept is that soil without a living component is simply a part of earth's outer layer.

Further refinement of the soil concept is occurring in view of an appreciation of energy transport and transformation within soil. The term is popularly applied to the material on the surface of the Earth's moon and Mars, a usage acceptable within a portion of the scientific community. Accurate to this modern understanding of soil is Nikiforoff's 1959 definition of soil as the "excited skin of the sub aerial part of the earth's crust".

Areas of Practice

Academically, soil scientists tend to be drawn to one of five areas of specialization: microbiology, pedology, edaphology, physics or chemistry. Yet the work specifics are very much dictated by the challenges facing our civilization's desire to sustain the land that supports it, and the distinctions between the sub-disciplines of soil science often blur in the process. Soil science professionals commonly stay current in soil chemistry, soil physics, soil microbiology, pedology, and applied soil science in related disciplines

One interesting effort drawing in soil scientists in the USA as of 2004 is the Soil Quality Initiative. Central to the Soil Quality Initiative is developing indices of soil health and then monitoring them in a way that gives us long term (decade-to-decade) feedback on our performance as stewards of the planet. The effort includes understanding the functions of soil microbiotic crusts and exploring the potential to sequester atmospheric carbon in soil organic matter. The concept of soil quality, however, has not been without its share of controversy and criticism, including critiques by Nobel Laureate Norman Borlaug and World Food Prize Winner Pedro Sanchez.

A more traditional role for soil scientists has been to map soils. Most every area in the United States now has a published soil survey, which includes interpretive tables as to how soil properties support or limit activities and uses. An internationally accepted soil taxonomy allows uniform communication of soil characteristics and functions. National and international soil survey efforts have given the profession unique insights into landscape scale functions.

EDAPHOLOGY

Edaphology is one of two main divisions of soil science, the other being pedology. Edaphology is concerned with the influence of soils on living things, particularly plants. The term is also applied to the study of how soil influences man's use of land for plant growth as well as man's overall use of the land. General subfields within edaphology are agricultural soil science (known by the term agrology in some regions) and environmental soil science. In Russia, edaphology is considered equivalent to pedology, but is recognized to have an applied sense consistent with agrophysics and agrochemistry outside of Russia.

History

Xenophon (431–355 BC), and Cato (234–149 BC), were early edaphologists. Xenophon noted the beneficial effect of turning a cover crop into the earth. Cato wrote De Agri Cultura ("On Farming") which recommended tillage, crop rotation and the use of legumes in the rotation to build soil nitrogen.

He also devised the first soil capability classification for specific crops. Jan Baptist van Helmont (1577–1644) performed a famous experiment, growing a willow tree in a pot of soil and supplying only rainwater for five years. The weight gained by the tree was greater than the weight loss of the soil. He concluded that the willow was made of water. Although only partly correct, his experiment reignited interest in edaphology.

AREAS OF STUDY

Agricultural Soil Science

Agricultural soil science is the application of soil chemistry, physics, and biology dealing with the production of crops. In terms of soil chemistry, it places particular emphasis on plant nutrients of importance to farming and horticulture, especially with regard to soil fertility and fertilizer components. Physical edaphology is strongly associated with crop irrigation and drainage.

Soil husbandry is a strong tradition within agricultural soil science. Beyond preventing soil erosion and degradation in cropland, soil husbandry seeks to sustain the agricultural soil resource though the use of soil conditioners and cover crops.

Environmental Soil Science

Environmental soil science studies our interaction with the pedosphere on beyond crop production. Fundamental and applied aspects the field address vadose zone functions, septic drain field site assessment and function, land treatment of wastewater, stormwater, erosion control, soil contamination with metals and pesticides, remediation of contaminated soils, restoration of wetlands, soil degradation, and environmental nutrient management. It also studies soil in the context of land use planning, global warming, and acid rain.

PEDOLOGY

Pedology is the study of soils in their natural environment. It is one of two main branches of soil science, the other being edaphology. Pedology deals with pedogenesis, soil morphology, and soil classification, while edaphology studies the way soils influence plants, fungi, and other living things.

Overview

Soil is not only a support for vegetation, but it is also the zone (the pedosphere) of numerous interactions between climate (water, air, temperature), soil life (micro-organisms, plants, animals) and its residues, the mineral material of the original and added rock, and its position in the landscape. During its formation and genesis, the soil profile slowly deepens and develops characteristic layers, called 'horizons', while a steady state balance is approached.

Soil users (such as agronomists) showed initially little concern in the dynamics of soil. They saw it as medium whose chemical, physical and biological properties were useful for the services of agronomic productivity. On the other hand, pedologists and geologists did not initially focus on the agronomic applications of the soil characteristics (edaphic properties) but upon its relation to the nature and history of landscapes. Today, there's an integration of the two disciplinary approaches as part of landscape and environmental sciences.

Pedologists are now also interested in the practical applications of a good understanding of pedogenesis processes (the evolution and functioning of soils), like interpreting its environmental history and predicting consequences of changes in land use, while agronomists understand that the cultivated soil is a complex medium, often resulting from several thousands of years of evolution. They understand that the current balance is fragile and that only a thorough knowledge of its history makes it possible to ensure its sustainable use.

Concepts

- Complexity in soil genesis is more common than simplicity.
- Soils lie at the interface of Earth's atmosphere, biosphere, hydrosphere and lithosphere. Therefore, a thorough understanding of soils requires some knowledge of meteorology, climatology, ecology, biology, hydrology, geomorphology, geology and many other earth sciences and natural sciences.
- Contemporary soils carry imprints of pedogenic processes that were active in the past, although in many cases these imprints are difficult to observe or quantify. Thus, knowledge of paleoecology, palaeogeography, glacial geology and paleoclimatology is important for the recognition and understanding of soil genesis and constitute a basis for predicting the future soil changes.
- Five major, external factors of soil formation (climate, organisms, relief, parent material and time), and several smaller, less identifiable ones, drive pedogenic processes and create soil patterns.
- Distinctive bioclimatic regimes or combinations of pedogenic processes produce distinctive soils. Thus, distinctive, observable morphological features, e.g., illuvial clay accumulation in B horizons, are produced by certain combinations of pedogenic processes operative over varying periods of time.
- Pedogenic (soil-forming) processes act to both create and destroy order (anisotropy) within soils; these processes can proceed simultaneously. The resulting soil profile reflects the balance of these processes, present and past.
- The geological Principle of Uniformitarianism applies to soils, i.e., pedogenic processes active in soils today

have been operating for long periods of time, back to the time of appearance of organisms on the land surface. These processes do, however, have varying degrees of expression and intensity over space and time.

- A succession of different soils may have developed, eroded and/or regressed at any particular site, as soil genetic factors and site factors, e.g., vegetation, sedimentation, geomorphology, change.
- There are very few old soils (in a geological sense) because they can be destroyed or buried by geological events, or modified by shifts in climate by virtue of their vulnerable position at the surface of the earth. Little of the soil continuum dates back beyond the Tertiary period and most soils and land surfaces are no older than the Pleistocene Epoch. However, preserved/lithified soils (paleosols) are an almost ubiquitous feature in terrestrial (land-based) environments throughout most of geologic time. Since they record evidence of ancient climate change, they present immense utility in understanding climate evolution throughout geologic history.
- Knowledge and understanding of the genesis of a soil is important in its classification and mapping.
- Soil classification systems cannot be based entirely on perceptions of genesis, however, because genetic processes are seldom observed and because pedogenic processes change over time.
- Knowledge of soil genesis is imperative and basic to soil use and management. Human influence on, or adjustment to, the factors and processes of soil formation can be best controlled and planned using knowledge about soil genesis.
- Soils are natural clay factories (clay includes both clay mineral structures and particles less than 2 μm in diameter). Shales worldwide are, to a considerable extent, simply soil clays that have been formed in the pedosphere and eroded and deposited in the ocean basins, to become lithified at a later date.

7

The Weather

Weather is the state of the atmosphere, to the degree that it is hot or cold, wet or dry, calm or stormy, clear or cloudy. Most weather phenomena occur in the troposphere, just below the stratosphere. Weather refers, generally, to day-to-day temperature and precipitation activity, whereas climate is the term for the average atmospheric conditions over longer periods of time. When used without qualification, "weather" is understood to be the weather of Earth.

Weather occurs due to density (temperature and moisture) differences between one place and another. These differences can occur due to the sun angle at any particular spot, which varies by latitude from the tropics. The strong temperature contrast between polar and tropical air gives rise to the jet stream. Weather systems in the mid-latitudes, such as extratropical cyclones, are caused by instabilities of the jet stream flow. Because the Earth's axis is tilted relative to its orbital plane, sunlight is incident at different angles at different times of the year. On Earth's surface, temperatures usually range ±40 °C (100 °F to "40 °F) annually. Over thousands of years, changes in Earth's orbit affect the amount and distribution of solar energy received by the Earth and influence long-term climate

Surface temperature differences in turn cause pressure differences. Higher altitudes are cooler than lower altitudes due to differences in compressional heating. Weather forecasting is the application of science and technology to predict the state of the atmosphere for a future time and a given location. The atmosphere is a chaotic system, so small changes to one part of the system can grow to have large effects on the system as a whole. Human attempts to control the weather have occurred throughout human history, and there is evidence that human activity such as agriculture and industry has inadvertently modified weather patterns.

Studying how the weather works on other planets has been helpful in understanding how weather works on Earth. A famous landmark in the Solar System, Jupiter's *Great Red Spot*, is an anticyclonic storm known to have existed for at least 300 years. However, weather is not limited to planetary bodies. A star's corona is constantly being lost to space, creating what is essentially a very thin atmosphere throughout the Solar System. The movement of mass ejected from the Sun is known as the solar wind.

Cause

On Earth, common weather phenomena include wind, cloud, rain, snow, fog and dust storms. Less common events include natural disasters such as tornadoes, hurricanes, typhoons and ice storms. Almost all familiar weather phenomena occur in the troposphere (the lower part of the atmosphere). Weather does occur in the stratosphere and can affect weather lower down in the troposphere, but the exact mechanisms are poorly understood.

Weather occurs primarily due to density (temperature and moisture) differences between one place to another. These differences can occur due to the sun angle at any particular spot, which varies by latitude from the tropics. In other words, the farther from the tropics you lie, the lower the sun angle is, which causes those locations to be cooler due to the indirect sunlight. The strong temperature contrast between polar and tropical air gives rise to the jet stream. Weather systems in the mid-latitudes, such as extratropical cyclones, are caused by instabilities of the jet stream flow. Weather systems in the tropics, such as monsoons or organized thunderstorm systems, are caused by different processes. Because the Earth's axis is tilted relative to its orbital plane, sunlight is incident at different angles at different times of the year. In June the Northern Hemisphere is tilted towards the sun, so at any given Northern

Hemisphere latitude sunlight falls more directly on that spot than in December. This effect causes seasons. Over thousands to hundreds of thousands of years, changes in Earth's orbital parameters affect the amount and distribution of solar energy received by the Earth and influence long-term climate.

The uneven solar heating (the formation of zones of temperature and moisture gradients, or frontogenesis) can also be due to the weather itself in the form of cloudiness and precipitation. Higher altitudes are cooler than lower altitudes, which is explained by the lapse rate. On local scales, temperature differences can occur because different surfaces (such as oceans, forests, ice sheets, or man-made objects) have differing physical characteristics such as reflectivity, roughness, or moisture content. Surface temperature differences in turn cause pressure differences. A hot surface heats the air above it and the air expands, lowering the air pressure and its density. The resulting horizontal pressure gradient accelerates the air from high to low pressure, creating wind, and Earth's rotation then causes curvature of the flow via the Coriolis effect. The simple systems thus formed can then display emergent behaviour to produce more complex systems and thus other weather phenomena. Large scale examples include the Hadley cell while a smaller scale example would be coastal breezes.

The atmosphere is a chaotic system, so small changes to one part of the system can grow to have large effects on the system as a whole. This makes it difficult to accurately predict weather more than a few days in advance, though weather forecasters are continually working to extend this limit through the scientific study of weather, meteorology. It is theoretically impossible to make useful day-to-day predictions more than about two weeks ahead, imposing an upper limit to potential for improved prediction skill.

Chaos theory says that the slightest variation in the motion of the ground can grow with time. This idea is sometimes called the butterfly effect, from the idea that the motions caused by the flapping wings of a butterfly eventually could produce marked changes in the state of the atmosphere. Because of this sensitivity to small changes, it will never be possible to make perfect forecasts.

SHAPING THE PLANET EARTH

Weather is one of the fundamental processes that shape the Earth. The process of weathering breaks down rocks and soils into smaller fragments and then into their constituent substances. These are then free to take part in chemical reactions that can affect the surface further (such as acid rain) or are reformed into other rocks and soils. In this way, weather plays a major role in erosion of the surface.

EFFECT ON HUMANS

Effects on Populations

Weather has played a large and sometimes direct part in human history. Aside from climatic changes that have caused the gradual drift of populations (for example the desertification of the Middle East, and the formation of land bridges during glacial periods), extreme weather events have caused smaller scale population movements and intruded directly in historical events. One such event is the saving of Japan from invasion by the Mongol fleet of Kublai Khan by the Kamikaze winds in 1281. French claims to Florida came to an end in 1565 when a hurricane destroyed the French fleet, allowing Spain to conquer Fort Caroline. More recently, Hurricane Katrina redistributed over one million people from the central Gulf coast elsewhere across the United States, becoming the largest diaspora in the history of the United States. The Little Ice Age led to crop failure and great famines in Europe. In Finland, for example, the famine of 1696-97 killed one third of the population.

Effects on Individuals

Though weather affects people in drastic ways, it can also affect the human race in simpler ways. The human body is negatively affected by extremes in temperature, humidity, and wind.

Forecasting

Weather forecasting is the application of science and technology to predict the state of the atmosphere for a future time and a given location. Human beings have attempted to predict the weather informally for millennia, and formally since at least the nineteenth century. Weather forecasts are made by collecting quantitative data about the current state of the atmosphere and using scientific understanding of atmospheric processes to project how the atmosphere will evolve.

Once an all-human endeavor based mainly upon changes in barometric pressure, current weather conditions, and sky condition, forecast models are now used to determine future conditions. Human input is still required to pick the best possible forecast model to base the forecast upon, which involves pattern recognition skills, teleconnections, knowledge of model performance, and knowledge of model biases. The chaotic nature of the atmosphere, the massive computational power required to solve the equations that describe the atmosphere, error involved in measuring the initial conditions, and an incomplete understanding of atmospheric processes mean that forecasts become less accurate as the difference in current time and the time for which the forecast is being made (the *range* of the forecast) increases. The use of ensembles and model consensus helps to narrow the error and pick the most likely outcome.

There are a variety of end users to weather forecasts. Weather warnings are important forecasts because they are used to protect life and property. Forecasts based on temperature and precipitation are important to agriculture, and therefore to commodity traders within stock markets. Temperature forecasts are used by utility companies to estimate demand over coming days. On an everyday basis, people use weather forecasts to determine what to wear on a given day. Since outdoor activities are severely curtailed by heavy rain, snow and the wind chill, forecasts can be used to plan activities around these events, and to plan ahead and survive them.

Modification

The aspiration to control the weather is evident throughout human history: from ancient rituals intended to bring rain for crops to the U. S. Military Operation Popeye, an attempt to disrupt supply lines by lengthening the North Vietnamese monsoon. The most successful attempts at influencing weather involve cloud seeding; they include the fog-and low stratus dispersion techniques employed by major airports, techniques used to increase winter precipitation over mountains, and techniques to suppress hail. A recent example of weather control was China's preparation for the 2008 Summer Olympic Games. China shot 1,104 rain dispersal rockets from 21 sites in the city of Beijing in an effort to keep rain away from the opening ceremony of the games on August 8, 2008. Guo Hu, head of the Beijing Municipal Meteorological Bureau (BMB), confirmed the success of the operation with 100 millimeters falling in Baoding City of Hebei Province, to the southwest and Beijing's Fangshan District recording a rainfall of 25 millimeters.

Whereas there is inconclusive evidence for these techniques' efficacy, there is extensive evidence that human activity such as agriculture and industry results in inadvertent weather modification:

- Acid rain, caused by industrial emission of sulfur dioxide and nitrogen oxides into the atmosphere, adversely effects freshwater lakes, vegetation, and structures.
- Anthropogenic pollutants reduce air quality and visibility.
- Climate change caused by human activities that emit greenhouse gases into the air is expected to affect the frequency of extreme weather events such as drought, extreme temperatures, flooding, high winds, and severe storms.

The effects of inadvertent weather modification may pose serious threats to many aspects of civilization, including ecosystems, natural resources, food and fiber production, economic development, and human health.

EXTREMES ON EARTH

On Earth, temperatures usually range ±40 °C (100 °F to "40 °F) annually. The range of climates and latitudes across the planet can offer extremes of temperature outside this range. The coldest air temperature ever recorded on Earth is "89.2 °C ("128.6 °F), at Vostok Station, Antarctica on 21 July 1983. The hottest air temperature ever recorded was 57.7 °C (135.9 °F) at Al 'Aziziyah, Libya, on September 13, 1922, but that reading is queried. The highest recorded average annual temperature was 34.4 °C (93.9 °F) at Dallol, Ethiopia. The coldest recorded average annual temperature was "55.1 °C ("67.2 °F) at Vostok Station, Antarctica. The coldest average annual temperature in a permanently inhabited location is at Eureka, Nunavut, in Canada, where the annual average temperature is “19.7 °C (“3.5 °F).

EXTRATERRESTRIAL WITHIN THE SOLAR SYSTEM

Studying how the weather works on other planets has been seen as helpful in understanding how it works on Earth. Weather on other planets follows many of the same physical principles as weather on Earth, but occurs on different scales and in atmospheres having different chemical composition. The Cassini–Huygens mission to Titan discovered clouds formed from methane or ethane which deposit rain composed of liquid methane and other organic compounds. Earth's atmosphere includes six latitudinal circulation zones, three in each hemisphere. In contrast, Jupiter's banded appearance shows many such zones, Titan has a single jet stream near the 50th parallel north latitude, and Venus has a single jet near the equator.

One of the most famous landmarks in the Solar System, Jupiter's *Great Red Spot*, is an anticyclonic storm known to have existed for at least 300 years. On other gas giants, the lack of a surface allows the wind to reach enormous speeds: gusts of up to 600 metres per second (about 2,100 km/h or 1,300 mph) have been measured on the planet Neptune. This has created a puzzle for planetary scientists. The weather is ultimately created by solar energy and the amount of energy received by Neptune is only about D $_{900}$ of that received by Earth, yet the intensity of weather phenomena on Neptune is far greater than on Earth. The strongest planetary winds discovered so far are on the extrasolar planet HD 189733 b, which is thought to have easterly winds moving at more than 9,600 kilometres per hour (6,000 mph).

Space Weather

Weather is not limited to planetary bodies. Like all stars, the sun's corona is constantly being lost to space, creating what is essentially a very thin atmosphere throughout the Solar System. The movement of mass ejected from the Sun is known as the solar wind. Inconsistencies in this wind and larger events on the surface of the star, such as coronal mass ejections, form a system that has features analogous to conventional weather systems (such as pressure and wind) and is generally known as space weather. Coronal mass ejections have been tracked as far out in the solar system as Saturn. The activity of this system can affect planetary atmospheres and occasionally surfaces. The interaction of the solar wind with the terrestrial atmosphere can produce spectacular aurorae, and can play havoc with electrically sensitive systems such as electricity grids and radio signals.

WEATHER STATION

A weather station is a facility with instruments and equipment for observing atmospheric conditions to provide information for weather forecasts and to study the weather and climate. The measurements taken include temperature, barometric pressure, humidity, wind speed, wind direction, and precipitation amounts. Wind measurements are taken as free of other obstructions as possible, while temperature and humidity measurements are kept free from direct solar radiation, or insolation. Manual observations are taken at least once daily, while automated observations are taken at least once an hour.

Instruments

Typical weather stations have the following instruments:

- Thermometer for measuring temperature
- Barometer for measuring barometric pressure/air pressure
- Hygrometer for measuring humidity
- Anemometer for measuring wind speed
- Wind vane for measuring wind direction
- Rain gauge for measuring precipitation.

In addition, at certain Automated airport weather stations, additional instruments may be employed, including:

- Present Weather/Precipitation Identification Sensor for identifying falling precipitation
- Disdrometer for measuring drop size distribution
- Transmissometer for measuring visibility
- Ceiling projector for measuring cloud ceiling.

More sophisticated stations may also measure the ultraviolet index, solar radiation, leaf wetness, soil moisture, soil temperature, water temperature in ponds, lakes, creeks, or rivers, and occasionally other data.

Exposure

Except for those instruments requiring direct exposure to the elements (anemometer, rain gauge), the instruments should be sheltered in a vented box, usually a Stevenson screen, to keep direct sunlight off the thermometer and wind off the hygrometer. The instrumentation may be specialized to allow for periodic recording otherwise significant manual labour is required for record keeping. Automatic transmission of data, in a format such as METAR, is also desirable as many weather station's data is required for weather forecasting.

Personal Weather Station

A personal weather station is a set of weather measuring instruments operated by a private individual, club, association, or even business (where obtaining and distributing weather data is not a part of the entity's business operation). The quality and number of instruments can vary widely, and placement of the instruments, so important to obtaining accurate, meaningful, and comparable data, can also be very variable.

Today's personal weather stations also typically involve a digital console that provides readouts of the data being collected. These consoles may interface to a personal computer where data can be displayed, stored, and uploaded to Web sites or data ingestion/distribution systems.

Personal weather stations may be operated solely for the enjoyment and education of the owner, but many personal weather station operators also share their data with others, either by manually compiling data and distributing it, or through use of the Internet or amateur radio. The Citizen Weather Observer Program (CWOP) is one such, and the data submitted through use of software, a personal computer, and internet connection (or amateur radio) are utilized by the National Weather Service when generating forecast models, and by many other entities as well. Each weather station submitting data to CWOP will also have an individual web page that depicts the data submitted by that station. The Weather Underground Internet site is another popular destination for the submittal and sharing of data with others around the world. As with CWOP, each station submitting data to The Weather Underground has a unique web page displaying their submitted data.

WEATHER FORECASTING

Weather forecasting is the application of science and technology to predict the state of the atmosphere for a future time and a given location. Human beings have attempted to predict the weather informally for millennia, and formally since at least the nineteenth century. Weather forecasts are made by collecting quantitative data about the current state of the atmosphere and using scientific understanding of atmospheric processes to project how the atmosphere will evolve.

Once an all-human endeavor based mainly upon changes in barometric pressure, current weather conditions, and sky condition, forecast models are now used to determine future conditions. Human input is still required to pick the best possible forecast model to base the forecast upon, which involves pattern recognition skills, teleconnections, knowledge of model performance, and knowledge of model biases. The chaotic nature of the atmosphere, the massive computational power required to solve the equations that describe the atmosphere, error involved in measuring the initial conditions, and an incomplete understanding of atmospheric processes mean that forecasts become less accurate as the difference in current time and the time for which the forecast is being made (the *range* of the forecast) increases. The use of ensembles and model consensus help narrow the error and pick the most likely outcome.

There are a variety of end uses to weather forecasts. Weather warnings are important forecasts because they are used to protect life and property. Forecasts based on temperature and precipitation are important to agriculture, and therefore to traders within commodity markets. Temperature forecasts are used by utility companies to estimate demand over coming days. On an everyday basis, people use weather forecasts to determine what to wear on a given day. Since outdoor activities are severely curtailed by heavy rain, snow and the wind chill, forecasts can be used to plan activities around these events, and to plan ahead and survive them.

History

For millennia people have tried to forecast the weather. In 650 BC, the Babylonians predicted the weather from cloud patterns as well as astrology. In about 340 BC, Aristotle described weather patterns in *Meteorologica*. Later, Theophrastus compiled a book on weather forecasting, called the *Book of Signs*. Chinese weather prediction lore extends at least as far back as 300 BC, which was also around the same time ancient Indian astronomers developed weather-prediction methods. In 904 AD, Ibn Wahshiyya's *Nabatean Agriculture* discussed the weather forecasting of atmospheric changes and signs from the planetary astral alterations; signs of rain based on observation of the lunar phases; and weather forecasts based on the movement of winds.

Ancient weather forecasting methods usually relied on observed patterns of events, also termed pattern recognition. For example, it might be observed that if the sunset was particularly red, the following day often brought fair weather. This experience accumulated over the generations to produce weather lore. However, not all of these predictions prove reliable, and many of them have since been found not to stand up to rigorous statistical testing.

It was not until the invention of the electric telegraph in 1835 that the modern age of weather forecasting began. Before this time, it was not widely practicable to transport information about the current state of the weather any faster than a steam train (and the train also was a very new technology at that time). By the late 1840s, the telegraph allowed reports of weather conditions from a wide area to be received almost instantaneously, allowing forecasts to be made from knowledge of weather conditions further upwind. The two men most credited with the birth of forecasting as a science were Francis Beaufort (remembered chiefly for the Beaufort scale) and his protégé Robert Fitzroy (developer of the Fitzroy barometer). Both were influential men in British naval and governmental circles, and though ridiculed in the press at the time, their work gained scientific credence, was accepted by the Royal Navy, and formed the basis for all of today's weather forecasting knowledge. To convey information accurately, it became necessary to have a standard vocabulary describing clouds; this was achieved by means of a series of classifications and, in the 1890s, by pictorial cloud atlases.

Great progress was made in the science of meteorology during the 20th century. The possibility of numerical weather prediction was proposed by Lewis Fry Richardson in 1922, though computers did not exist to complete the vast number of calculations required to produce a forecast before the event had occurred. Practical use of numerical weather prediction began in 1955, spurred by the development of programmable electronic computers. In the United States, the first public radio forecasts were made in 1925 by Edward B. "E.B." Rideout, on WEEI, the Edison Electric Illuminating station in Boston. Rideout came from the U.S. Weather Bureau, as did WBZ weather forecaster G. Harold Noyes in 1931. Television forecasts followed with Jimmie Fidler in Cincinnati in 1940 or 1947 on the DuMont Television Network. The Weather Channel is a 24-hour cable network that began broadcasting in May 1982.

HOW MODELS CREATE FORECASTS

Data Collection

Surface weather observations of atmospheric pressure, temperature, wind speed and direction, humidity, precipitation are made near the Earth's surface by trained observers, automatic weather stations or buoys. The World Meteorological Organization acts to standardize the instrumentation, observing practices and timing of these observations worldwide. Stations either report hourly in METAR reports, or every six hours in SYNOP reports.

Measurements of temperature, humidity and wind above the surface are found by launching radiosondes on weather balloons. Data are usually obtained from near the surface to the middle of the stratosphere, about 21 km (13 mi). In recent years, data transmitted from commercial airplanes through the Aircraft Meteorological Data Relay (AMDAR) system has also been incorporated into upper air observation, primarily in numerical models.

Increasingly, data from weather satellites are being used because of their almost global coverage. Although their visible light images are very useful for forecasters to see development of clouds, little of this information can be used by numerical weather prediction models. The infrared (IR) data however can be used as it gives information on the temperature at the surface and cloud tops. Individual clouds can also be tracked from one time to the next to provide information on wind direction and strength at the clouds steering level. Both polar orbiting and geostationary satellites provide soundings of temperature and moisture throughout the depth of the atmosphere. Compared with similar data from radiosondes, the satellite data has the advantage of global coverage, however at a lower accuracy and resolution. Meteorological radar provide information on precipitation location and intensity, which can be used to estimate precipitation accumulations over time. Additionally, if a Pulse Doppler weather radar is used then wind speed and direction can be determined.

Data Assimilation and Analysis

During the data assimilation process, information gained from the observations is used in conjunction with a numerical model's most recent forecast for the time that observations were made, since this contains information from previous observations. This is used to produce a three-dimensional representation of the temperature, moisture and wind called a meteorological analysis. This is the model's estimate of the current state of the atmosphere.

Numerical Weather Prediction

Numerical weather prediction models are computer simulations of the atmosphere. They take the analysis as the starting point and evolve the state of the atmosphere forward in time using physics and fluid dynamics. The complicated equations which govern how the state of a fluid changes with time require supercomputers to solve them. The output from the model provides the basis of the weather forecast.

Model Output Post Processing

The raw output is often modified before being presented as the forecast. This can be in the form of statistical techniques to remove known biases in the model, or of adjustment to take into account consensus among other numerical weather forecasts. MOS or model output statistics is a technique used to interpret numerical model output and produce site-specific guidance. This guidance is presented in coded numerical form, and can be obtained for nearly all National Weather Service reporting stations in the United States.

Techniques

Persistence

The simplest method of forecasting the weather, persistence, relies upon today's conditions to forecast the conditions tomorrow. This can be a valid way of forecasting the weather when it is steady state, such as during the summer season in the tropics. This method of forecasting strongly depends upon the presence of a stagnant weather pattern. It can be useful in both short range forecasts and long range forecasts.

Use of a Barometer

Using barometric pressure and the pressure tendency (the change of pressure over time) has been used in forecasting since the late 19th century. The larger the change in pressure, especially if more than 3.5 hPa (0.10 inHg), the larger the change in weather can be expected. If the pressure drop is rapid, a low pressure system is approaching, and there is a greater chance of rain. Rapid pressure rises are associated with improving weather conditions, such as clearing skies.

Looking at the Sky

Along with pressure tendency, use of the sky condition is one of more important weather parameters that can be used to forecast weather in mountainous areas. Thickening of cloud cover or the invasion of a higher cloud deck is indicative of rain in the near future. Morning fog portends fair conditions, as rainy conditions are preceded by wind or clouds which prevent fog formation. The approach of a line of thunderstorms could indicate the approach of a cold front. Cloud-free skies are indicative of fair weather for the near future. The use of sky cover in weather prediction has led to various weather lore over the centuries.

Nowcasting

The forecasting of the weather within the next six hours is often referred to as nowcasting. In this time range it is possible to forecast smaller features such as individual showers and thunderstorms with reasonable accuracy, as well as other features too small to be resolved by a computer model. A human given the latest radar, satellite and observational data will be able to make a better analysis of the small scale features present and so will be able to make a more accurate forecast for the following few hours.

Use of Forecast Models

In the past, the human forecaster was responsible for generating the entire weather forecast based upon available observations. Today, human input is generally confined to choosing a model based on various parameters, such as model biases and performance. Using a consensus of forecast models, as well as ensemble members of the various models, can help reduce forecast error. However, regardless how small the average error becomes with any individual system, large errors within any particular piece of guidance are still possible on any given model run. Humans are required to interpret the model data into weather forecasts that are understandable to the end user. Humans can use knowledge of local effects which may be too small in size to be resolved by the model to add information to the forecast. While increasing accuracy of forecast models implies that humans may no longer be needed in the forecast process at some point in the future, there is currently still a need for human intervention.

Analog Technique

The Analog technique is a complex way of making a forecast, requiring the forecaster to remember a previous weather event which is expected to be mimicked by an upcoming event. What makes it a difficult technique to use is that there is rarely a perfect analog for an event in the future. Some call this type of forecasting pattern recognition. It remains a useful method of observing rainfall over data voids such as oceans, as well as the forecasting of precipitation amounts and distribution in the future. A similar technique is used in medium range forecasting, which is known as teleconnections, when systems in other locations are used to help pin down the location of another system within the surrounding regime. An example of teleconnections are by using El Niño-Southern Oscillation (ENSO) related phenomena.

- Analog model – A model based on similarities between the system under study and another system or process.
- Analytical model – A model that uses classic methods such as calculus or algebra to solve a series of equations.
- Conceptual model – A simplified representation of the system being examined.
- Continuous model – A model that uses continuous simulation, as opposed to a single-event model.
- Deterministic model – A model that produces the same output for a given input without consideration for risk or uncertainty.
- Empirical model – A model represented by simplified processes based on observation, measurements, or practical experience rather than solely on principles or theory. A lumped model is an example.
- Explicit model – A numerical model that uses parameter values or unknown variables at the beginning of a time step in the computational algorithms.
- Implicit model – A numerical model that uses parameter values or unknown variables at the end of a time step in the computational algorithms.
- Mass balance model – A model based on the conservation of mass and focuses on balancing inputs and outputs from the model area. Also known as a zero-dimensional model.
- Numerical model – A model that uses a numerical method to solve a series of equations, as opposed to an analytical model. The results from numerical models are often approximations, while analytic models produce exact solutions.
- One-dimensional model – A model that includes only one space dimension.
- Pseudo-deterministic model – A semi-distributed model.

- Stochastic mathematical model – A model that includes statistical elements and produces a set of outputs for a given set of inputs. The output represents a set of expected values.
- Two-dimensional model – A model that includes two space dimensions, usually horizontal and vertical averaging.

Public Uses

Most end users of forecasts are members of the general public. Thunderstorms can create strong winds and dangerous lightning strikes that can lead to deaths, power outages, and widespread hail damage. Heavy snow or rain can bring transportation and commerce to a stand-still, as well as cause flooding in low-lying areas. Excessive heat or cold waves can sicken or kill those with inadequate utilities, and droughts can impact water usage and destroy vegetation.

Several countries employ government agencies to provide forecasts and watches/warnings/advisories to the public in order to protect life and property and maintain commercial interests. Knowledge of what the end user needs from a weather forecast must be taken into account to present the information in a useful and understandable way. Examples include the National Oceanic and Atmospheric Administration's National Weather Service (NWS) and Environment Canada's Meteorological Service (MSC). Traditionally, newspaper, television, and radio have been the primary outlets for presenting weather forecast information to the public. Increasingly, the internet is being used due to the vast amount of specific information that can be found. In all cases, these outlets update their forecasts on a regular basis.

Severe Weather Alerts and Advisories

A major part of modern weather forecasting is the severe weather alerts and advisories which the national weather services issue in the case that severe or hazardous weather is expected. This is done to protect life and property. Some of the most commonly known of severe weather advisories are the severe thunderstorm and tornado warning, as well as the severe thunderstorm and tornado watch. Other forms of these advisories include winter weather, high wind, flood, tropical cyclone, and fog. Severe weather advisories and alerts are broadcast through the media, including radio, using emergency systems as the Emergency Alert System which break into regular programming.

Air Traffic

Because the aviation industry is especially sensitive to the weather, accurate weather forecasting is essential. Fog or exceptionally low ceilings can prevent many aircraft from landing and taking off. Turbulence and icing are also significant in-flight hazards. Thunderstorms are a problem for all aircraft because of severe turbulence due to their updrafts and outflow boundaries, icing due to the heavy precipitation, as well as large hail, strong winds, and lightning, all of which can cause severe damage to an aircraft in flight. Volcanic ash is also a significant problem for aviation, as aircraft can lose engine power within ash clouds.

On a day to day basis airliners are routed to take advantage of the jet stream tailwind to improve fuel efficiency. Aircrews are briefed prior to takeoff on the conditions to expect en route and at their destination. Additionally, airports often change which runway is being used to take advantage of a headwind. This reduces the distance required for takeoff, and eliminates potential crosswinds.

Marine

Commercial and recreational use of waterways can be limited significantly by wind direction and speed, wave periodicity and heights, tides, and precipitation. These factors can each influence the safety of marine transit. Consequently, a variety of codes have been established to efficiently transmit detailed marine weather forecasts to vessel pilots via radio, for example the MAFOR (marine forecast). Typical weather forecasts can be received at sea through the use of RTTY, Navtex and Radiofax.

Agriculture

Farmers rely on weather forecasts to decide what work to do on any particular day. For example, drying hay is only feasible in dry weather. Prolonged periods of dryness can ruin cotton, wheat, and corn crops. While corn crops can be ruined by drought, their dried remains can be used as a cattle feed substitute in the form of silage. Frosts and freezes play havoc with crops both during the spring and fall. For example, peach trees in full bloom can have their potential peach crop decimated by a spring freeze. Orange groves can suffer significant damage during frosts and freezes, regardless of their timing.

Forestry

Weather forecasting of wind, precipitations and humidity is essential for preventing and controlling wildfires. Different indices, like the *Forest fire weather index* and the *Haines Index*, have been developed to predict the areas more at risk to experience fire from natural or human causes. Conditions for the development of harmful insects can be predicted by forecasting the evolution of weather, too.

Utility Companies

Electricity and gas companies rely on weather forecasts to anticipate demand which can be strongly affected by the weather. They use the quantity termed the degree day to determine how strong of a use there will be for heating (heating degree day) or cooling (cooling degree day). These quantities are based on a daily average temperature of 65 °F (18 °C). Cooler temperatures force heating degree days (one per degree Fahrenheit), while warmer temperatures force cooling degree days. In winter, severe cold weather can cause a surge in demand as people turn up their heating. Similarly, in summer a surge in demand can be linked with the increased use of air conditioning systems in hot weather. By anticipating a surge in demand, utility companies can purchase additional supplies of power or natural gas before the price increases, or in some circumstances, supplies are restricted through the use of brownouts and blackouts.

Private Sector

Increasingly, private companies pay for weather forecasts tailored to their needs so that they can increase their profits or avoid large losses. For example, supermarket chains may change the stocks on their shelves in anticipation of different consumer spending habits in different weather conditions.

Weather forecasts can be used to invest in the commodity market, such as futures in oranges, corn, soybeans, and oil.

MILITARY APPLICATIONS

United States Armed Forces

Similarly to the private sector, military weather forecasters present weather conditions to the war fighter community. Military weather forecasters provide pre-flight and in-flight weather briefs to pilots and provide real time resource protection services for military installations.

Naval forecasters cover the waters and ship weather forecasts. The United States Navy provides a special service to both themselves and the rest of the federal government by issuing forecasts for tropical cyclone across the Pacific and Indian Oceans through their Joint Typhoon Warning Centre.

US Air Force

Within the United States, Air Force Weather provides weather forecasting for the Air Force and the Army. Air Force forecasters cover air operations in both wartime and peacetime operations and provide Army support; United States Coast Guard marine science technicians provide ship forecasts for ice breakers and other various operations within their realm; and Marine forecasters provide support for ground-and air-based United States Marine Corps operations.

All four military branches take their initial enlisted meteorology technical training at Keesler Air Force Base. Military and civilian forecasters actively cooperate in analyzing, creating and critiquing weather forecast products.

SPACE WEATHER

Space weather is the concept of changing environmental conditions in near-Earth space. It is distinct from the concept of weather within a planetary atmosphere, and deals with phenomena involving ambient plasma, magnetic fields, radiation and other matter in space. "Space weather" often implicitly means the conditions in near-Earth space within the magnetosphere and ionosphere, but it is also studied in interplanetary (and occasionally interstellar) space.

Within our own solar system, space weather is greatly influenced by the speed and density of the solar wind and the interplanetary magnetic field (IMF) carried by the solar wind plasma. A variety of physical phenomena are associated with space weather, including geomagnetic storms and substorms, energization of the Van Allen radiation belts, ionospheric disturbances and scintillation, aurora and geomagnetically induced currents at Earth's surface. Coronal mass ejections and their associated shock waves are also important drivers of space weather as they can compress the magnetosphere and trigger geomagnetic storms. Solar energetic particles, accelerated by coronal mass ejections or solar flares, are also an important driver of space weather as they can damage electronics onboard spacecraft through induced electric currents (e.g. Galaxy 15 failure), and threaten the life of astronauts.

Space weather exerts a profound influence in several areas related to space exploration and development. Changing geomagnetic conditions can induce changes in atmospheric density causing the rapid degradation of spacecraft altitude in low Earth orbit. Geomagnetic storms due to increased solar activity can potentially blind sensors aboard spacecraft, or interfere with on-board electronics. An understanding of space environmental conditions is also important in designing shielding and life support systems for manned spacecraft. There is also some concern that geomagnetic storms may also expose conventional aircraft flying at high altitudes to increased amounts of radiation.

Satellites Observing Space Weather

Since 1995, the joint NASA-ESA Solar and Heliospheric Observatory (SOHO) spacecraft is the main source of near-real time solar data for space weather prediction. It was joined in 1998 by the NASA Advanced Composition Explorer (ACE), which carries a space weather beacon for continuous transmission of relevant in situ space environment data. SOHO and ACE are located near the L_1 Lagrangian point, 1% of the earth-sun distance upstream of the earth where it measures solar wind plasma and magnetic field approximately one hour before it reaches the earth. Most recently, the launch of the NASA-ESA Solar-Terrestrial Relations Observatory (STEREO) added an additional space weather data stream that covers the region between the sun and the earth with stereoscopic imagery. The two STEREO spacecraft drift away from the earth by about 22 degrees per year, one leading and the other trailing the earth in its orbit.

Modeling Efforts

Major modelling efforts to simulate the environment from the Sun to the Earth and beyond using three-dimensional magnetohydrodynamics frameworks have been undertaken since the 1990s. In the United States, the two major centers are the Centre for Space Environment Modeling (CSEM) and the Centre for Integrated Space weather Modeling (CISM).

Examples of Space Weather Events

- September 2, 1859, disruption of telegraph service.
- The best-known example of space weather events is the collapse of the Hydro-Québec power network on March 13, 1989 due to geomagnetically induced currents. This was started by a transformer failure, which led to a general blackout, which lasted more than 9 hours and affected 6 million people. The geomagnetic storm causing this event was itself the result of a Coronal Mass Ejection, ejected from the Sun on March 9, 1989.
- A geomagnetic storm on January 20, 1994 temporarily knocked out two Canadian communications satellites, Aniks E1 and E2 and the international communication satellite Intelsat K.
- A Coronal Mass Ejection on January 7, 1997 hit the Earth's magnetosphere on January 10 and caused the loss of the AT&T Telstar 401 communication satellite (a $200 million value).
- Transpolar routes flown by airplanes are particularly sensitive to space weather, in part because of Federal Aviation Regulations requiring reliable communication over the entire flight. It is estimated to cost about $100,000 each time such a flight is diverted from a polar route. Nine airlines are currently operating polar routes. Receiver Autonomous Integrity Monitoring technology can help planes get accurate GPS signals

even when some satellite signals are experiencing interference.

- No large Solar Energetic Particles events have happened during a manned mission. However, such a large event happened on August 7, 1972, between the Apollo 16 and Apollo 17 lunar missions. The dose of particles which would have hit an astronaut outside of earth's protective magnetic field, had this event happened during one of these missions, would have been deadly or at least life-threatening.
- Nozomi Mars Probe was hit by a large Solar Energetic Particles event on April 21, 2002, which caused large-scale failure. The mission, which was already about 3 years behind schedule, was eventually abandoned in December 2003.

Space Weather at the Earth's Surface

Changes in the near-Earth space environment effect our society. The best known ground-level consequence of space weather is geomagnetically induced currents, or GIC. These are damaging electrical currents that can flow in power grids, pipelines and other conducting networks. Rapid magnetic changes on the ground-that occur during geomagnetic storms and are associated with space weather-can also be important for activities such as geophysical mapping and hydrocarbon production.

Geophysical Exploration

Air and ship borne magnetic surveys can be affected by rapid magnetic field variations during geomagnetic storms. Storms can cause data interpretation problems where the magnetic field changes due to space weather are of similar magnitude to those of the sub-surface crustal magnetic field in the survey area. Accurate geomagnetic storm warnings, including an assessment of the magnitude and duration of the storm, allows for an economic use of survey equipment.

Geophysics and Hydrocarbon Production

For economic and other reasons, oil and gas production often involves the directional drilling of well paths many kilometres from a single wellhead in both the horizontal and vertical directions. The accuracy requirements are strict, due to target size – reservoirs may only be a few tens to hundreds of meters across – and for safety reasons, because of the proximity of other boreholes. Surveying by the most accurate gyroscopic method is expensive, since it can involve the cessation of drilling for a number of hours. An alternative is to use a magnetic survey, which enables measurement while drilling (MWD). Near real time magnetic data can be used to correct the drilling direction and nearby magnetic observatories prove vital. Magnetic data and storm forecasts can also be helpful in clarifying unknown sources of drilling error on an on-going basis.

8

The Climate

Climates encompasses the statistics of temperature, humidity, atmospheric pressure, wind, rainfall, atmospheric particle count and other meteorological elements in a given region over long periods of time. Climate can be contrasted to weather, which is the present condition of these same elements and their variations over periods up to two weeks.

The climate of a location is affected by its latitude, terrain, and altitude, as well as nearby water bodies and their currents. Climates can be classified according to the average and the typical ranges of different variables, most commonly temperature and precipitation. The most commonly used classification scheme was originally developed by Wladimir Köppen.

The Thornthwaite system, in use since 1948, incorporates evapotranspiration in addition to temperature and precipitation information and is used in studying animal species diversity and potential impacts of climate changes. The Bergeron and Spatial Synoptic Classification systems focus on the origin of air masses that define the climate of a region.

Paleoclimatology is the study of ancient climates. Since direct observations of climate are not available before the 19th century, paleoclimates are inferred from *proxy variables* that include non-biotic evidence such as sediments found in lake beds and ice cores, and biotic evidence such as tree rings and coral. Climate models are mathematical models of past, present and future climates.

Definition

Climate (from Ancient Greek *klima*, meaning *inclination*) is commonly defined as the weather averaged over a long period of time. The standard averaging period is 30 years, but other periods may be used depending on the purpose. Climate also includes statistics other than the average, such as the magnitudes of day-to-day or year-to-year variations. The Intergovernmental Panel on Climate Change (IPCC) glossary definition is:

Climate in a narrow sense is usually defined as the "average weather," or more rigorously, as the statistical description in terms of the mean and variability of relevant quantities over a period of time ranging from months to thousands or millions of years. The classical period is 30 years, as defined by the World Meteorological Organization (WMO). These quantities are most often surface variables such as temperature, precipitation, and wind. Climate in a wider sense is the state, including a statistical description, of the climate system.

The difference between climate and weather is usefully summarized by the popular phrase "Climate is what you expect, weather is what you get." Over historical time spans there are a number of nearly constant variables that determine climate, including latitude, altitude, proportion of land to water, and proximity to oceans and mountains. These change only over periods of millions of years due to processes such as plate tectonics. Other climate determinants are more dynamic: for example, the thermohaline circulation of the ocean leads to a 5 °C (9 °F) warming of the northern Atlantic ocean compared to other ocean basins. Other ocean currents redistribute heat between land and water on a more regional scale. The density and type of vegetation coverage affects solar heat absorption, water retention, and rainfall on a regional level. Alterations in the quantity of atmospheric greenhouse gases determines the amount of solar energy retained by the planet, leading to global warming or global cooling. The variables which determine climate are numerous and the interactions complex, but there is general agreement that the broad outlines are understood, at least insofar as the determinants of historical climate change are concerned.

Climate Classification

There are several ways to classify climates into similar regimes. Originally, climes were defined in Ancient Greece to describe the weather depending upon a location's latitude. Modern climate classification methods can be broadly divided into *genetic* methods, which focus on the causes of climate, and *empiric* methods, which focus on the effects of climate. Examples of genetic classification include methods based on the relative frequency of different air mass types or locations within synoptic weather disturbances. Examples of empiric classifications include climate zones defined by plant hardiness, evapotranspiration, or more generally the Köppen climate classification which was originally designed to identify the climates associated with certain biomes. A common shortcoming of these classification schemes is that they produce distinct boundaries between the zones they define, rather than the gradual transition of climate properties more common in nature.

Bergeron and Spatial Synoptic

The most generic classification is that involving the concept of air masses. The Bergeron classification is the most widely accepted form of air mass classification. Air mass classification involves three letters. The first letter describes its moisture properties, with c used for continental air masses (dry) and m for maritime air masses (moist). The second letter describes the thermal characteristic of its source region: T for tropical, P for polar, A for Arctic or Antarctic, M for monsoon, E for equatorial, and S for superior air (dry air formed by significant downward motion in the atmosphere). The third letter is used to designate the stability of the atmosphere. If the air mass is colder than the ground below it, it is labeled k. If the air mass is warmer than the ground below it, it is labeled w. While air mass identification was originally used in weather forecasting during the 1950s, climatologists began to establish synoptic climatologies based on this idea in 1973.

Based upon the Bergeron classification scheme is the Spatial Synoptic Classification system (SSC). There are six categories within the SSC scheme: Dry Polar (similar to continental polar), Dry Moderate (similar to maritime superior), Dry Tropical (similar to continental tropical), Moist Polar (similar to maritime polar), Moist Moderate (a hybrid between maritime polar and maritime tropical), and Moist Tropical (similar to maritime tropical, maritime monsoon, or maritime equatorial).

Köppen

The Köppen classification depends on average monthly values of temperature and precipitation. The most commonly used form of the Köppen classification has five primary types labelled A through E. These primary types are A, tropical; B, dry; C, mild mid-latitude; D, cold mid-latitude; and E, polar. The five primary classifications can be further divided into secondary classifications such as rain forest, monsoon, tropical savanna, humid subtropical, humid continental, oceanic climate, Mediterranean climate, steppe, subarctic climate, tundra, polar ice cap, and desert.

Rain forests are characterized by high rainfall, with definitions setting minimum normal annual rainfall between 1,750 millimetres (69 in) and 2,000 millimetres (79 in). Mean monthly temperatures exceed 18 °C (64 °F) during all months of the year.

A monsoon is a seasonal prevailing wind which lasts for several months, ushering in a region's rainy season. Regions within North America, South America, Sub-Saharan Africa, Australia and East Asia are monsoon regimes. A tropical savanna is a grassland biome located in semi-arid to semi-humid climate regions of subtropical and tropical latitudes, with average temperatures remain at or above 18 °C (64 °F) year round and rainfall between 750 millimetres (30 in) and 1,270 millimetres (50 in) a year. They are widespread on Africa, and are also found in India, the northern parts of South America, Malaysia, and Australia.

The humid subtropical climate zone where winter rainfall (and sometimes snowfall) is associated with large storms that the westerlies steer from west to east. Most summer rainfall occurs during thunderstorms and from occasional tropical cyclones. Humid subtropical climates lie on the east side continents, roughly between latitudes 20° and 40° degrees away from the equator. A humid continental climate is marked by variable weather patterns and a large seasonal temperature variance. Places with more than three months of average daily temperatures above 10 °C (50 °F) and a coldest month temperature below "3 °C (26.6 °F) and which do not meet the criteria for an arid or semiarid climate, are classified as continental.

An oceanic climate is typically found along the west coasts at the middle latitudes of all the world's continents, and in southeastern Australia, and is accompanied by plentiful precipitation year round. The Mediterranean climate regime resembles the climate of the lands in the Mediterranean Basin, parts of western North America, parts of Western and South Australia, in southwestern South Africa and in parts of central Chile. The climate is characterized by hot, dry summers and cool, wet winters.

A steppe is a dry grassland with an annual temperature range in the summer of up to 40 °C (104 °F) and during the winter down to "40 °C ("40 °F). A subarctic climate has little precipitation, and monthly temperatures which are above 10 °C (50 °F) for one to three months of the year, with permafrost in large parts of the area due to the cold winters. Winters within subarctic climates usually include up to six months of temperatures averaging below 0 °C (32 °F).

Tundra occurs in the far Northern Hemisphere, north of the taiga belt, including vast areas of northern Russia and Canada. A polar ice cap, or polar ice sheet, is a high-latitude region of a planet or moon that is covered in ice. Ice caps form because high-latitude regions receive less energy in the form of solar radiation from the sun than equatorial regions, resulting in lower surface temperatures.

A desert is a landscape form or region that receives very little precipitation. Deserts usually have a large diurnal and seasonal temperature range, with high daytime temperatures (in summer up to 45 °C or 113 °F), and low night-time temperatures (in winter down to 0 °C; 32 °F) due to extremely low humidity. Many deserts are formed by rain shadows, as mountains block the path of moisture and precipitation to the desert.

Thornthwaite

Devised by the American climatologist and geographer C. W. Thornthwaite, this climate classification method monitors the soil water budget using the concept of evapotranspiration. It monitors the portion of total precipitation used to nourish vegetation over a certain area. It uses indices such as a humidity index and an aridity index to determine an area's moisture regime based upon its average temperature, average rainfall, and average vegetation type. The lower the value of the index in any given area, the drier the area is.

The moisture classification includes climatic classes with descriptors such as hyperhumid, humid, subhumid, subarid, semi-arid (values of-20 to-40), and arid (values below-40). Humid regions experience more precipitation than evaporation each year, while arid regions experience greater evaporation than precipitation on an annual basis. A total of 33 percent of the Earth's landmass is considered either arid of semi-arid, including southwest North America, southwest South America, most of northern and a small part of southern Africa, southwest and portions of eastern Asia, as well as much of Australia. Studies suggest that precipitation effectiveness (PE) within the Thornthwaite moisture index is overestimated in the summer and underestimated in the winter. This index can be effectively used to determine the number of herbivore and mammal species numbers within a given area. The index is also used in studies of climate change. Thermal classifications within the Thornthwaite scheme include microthermal, mesothermal, and megathermal regimes. A microthermal climate is one of low annual mean temperatures, generally between 0 °C (32 °F) and 14 °C (57 °F) which experiences short summers and has a potential evaporation between 14 centimetres (5.5 in) and 43 centimetres (17 in). A mesothermal climate lacks persistent heat or persistent cold, with potential evaporation between 57 centimetres (22 in) and 114 centimetres (45 in). A megathermal climate is one with persistent high temperatures and abundant rainfall, with potential annual evaporation in excess of 114 centimetres (45 in).

RECORD

Modern

Details of the modern climate record are known through the taking of measurements from such weather instruments as thermometers, barometers, and anemometers during the past few centuries. The instruments used to study weather conditions over the modern time scale, their known error, their immediate environment, and their exposure have changed over the years, which must be considered when studying the climate of centuries past.

Paleoclimatology

Paleoclimatology is the study of past climate over a great period of the Earth's history. It uses evidence from ice sheets, tree rings, sediments, coral, and rocks to determine the past state of the climate. It demonstrates periods of stability and periods of change and can indicate whether changes follow patterns such as regular cycles.

Climate Change

Climate change is the variation in global or regional climates over time. It reflects changes in the variability or average state of the atmosphere over time scales ranging from decades to millions of years. These changes can be caused by processes internal to the Earth, external forces (e.g. variations in sunlight intensity) or, more recently, human activities.

In recent usage, especially in the context of environmental policy, the term "climate change" often refers only to changes in modern climate, including the rise in average surface temperature known as global warming. In some cases, the term is also used with a presumption of human causation, as in the United Nations Framework Convention on Climate Change (UNFCCC). The UNFCCC uses "climate variability" for non-human caused variations.

Earth has undergone periodic climate shifts in the past, including four major ice ages. These consisting of glacial periods where conditions are colder than normal, separated by interglacial periods. The accumulation of snow and ice during a glacial period increases the surface albedo, reflecting more of the Sun's energy into space and maintaining a lower atmospheric temperature. Increases in greenhouse gases, such as by volcanic activity, can increase the global temperature and produce an interglacial. Suggested causes of ice age periods include the positions of the continents, variations in the Earth's orbit, changes in the solar output, and vulcanism.

Climate Models

Climate models use quantitative methods to simulate the interactions of the atmosphere, oceans, land surface and ice. They are used for a variety of purposes from study of the dynamics of the weather and climate system to projections of future climate. All climate models balance, or very nearly balance, incoming energy as short wave (including visible) electromagnetic radiation to the earth with outgoing energy as long wave (infrared) electromagnetic radiation from the earth. Any imbalance results in a change in the average temperature of the earth.

The most talked-about applications of these models in recent years have been their use to infer the consequences of increasing greenhouse gases in the atmosphere, primarily carbon dioxide. These models predict an upward trend in the global mean surface temperature, with the most rapid increase in temperature being projected for the higher latitudes of the Northern Hemisphere.

Models can range from relatively simple to quite complex:

- Simple radiant heat transfer model that treats the earth as a single point and averages outgoing energy
- this can be expanded vertically (radiative-convective models), or horizontally
- finally, (coupled) atmosphere–ocean–sea ice global climate models discretise and solve the full equations for mass and energy transfer and radiant exchange.

ATTRIBUTION OF RECENT CLIMATE CHANGE

Attribution of recent climate change is the effort to scientifically ascertain mechanisms responsible for recent changes observed in the Earth's climate. The effort has focused on changes observed during the period of instrumental temperature record, when records are most reliable; particularly on the last 50 years, when human activity has grown fastest and observations of the troposphere have become available. The dominant mechanisms to which recent climate change has been attributed all result from human activity. They are:

- increasing atmospheric concentrations of greenhouse gases

- global changes to land surface, such as deforestation
- increasing atmospheric concentrations of aerosols.

Attribution of recent change to anthropogenic forcing is based on the following facts:

- The observed change is not consistent with natural variability.
- Known natural forcings would, if anything, be negative over this period.
- Known anthropogenic forcings are consistent with the observed response.
- The pattern of the observed change is consistent with the anthropogenic forcing.

Recent reports from the Intergovernmental Panel on Climate Change (IPCC) have concluded that:

- "Most of the observed increase in globally averaged temperatures since the mid-20th century is *very likely* due to the observed increase in anthropogenic greenhouse gas concentrations."; *It is extremely unlikely (<5%) that the global pattern of warming during the past half century can be explained without external forcing* (i.e., it is inconsistent with being the result of internal variability), and very unlikely that it is due to known natural external causes alone. The warming occurred in both the ocean and the atmosphere and took place at a time when natural external forcing factors would likely have produced cooling.
- "From new estimates of the combined anthropogenic forcing due to greenhouse gases, aerosols, and land surface changes, it is *extremely likely* that human activities have exerted a substantial net warming influence on climate since 1750."
- "It is *virtually certain* that anthropogenic aerosols produce a net negative radiative forcing (cooling influence) with a greater magnitude in the Northern Hemisphere than in the Southern Hemisphere.

The panel defines "very likely," "extremely likely," and "virtually certain" as indicating probabilities greater than 90%, 95%, and 99%, respectively.

KEY ATTRIBUTIONS

Greenhouse Gases

Scientific consensus has identified carbon dioxide as the dominant greenhouse gas forcing. (The dominant greenhouse gas overall is water vapour. Water vapour, however, has a very short atmospheric lifetime (about 10 days) and is very nearly in a dynamic equilibrium in the atmosphere, so it is not a forcing gas in the context of global warming.) Methane and nitrous oxide are also major forcing contributors to the greenhouse effect. The Kyoto Protocol lists these together with hydrofluorocarbons (HFCs), perfluorocarbons (PFCs), and sulphur hexafluoride (SF6), which are entirely artificial (i.e. anthropogenic) gases which also contribute to radiative forcing in the atmosphere. The chart at right attributes anthropogenic greenhouse gas emissions to eight main economic sectors, of which the largest contributors are power stations (many of which burn coal or other fossil fuels), industrial processes (among which cement production is a dominant contributor), transportation fuels (generally fossil fuels), and agricultural by-products (mainly methane from enteric fermentation and nitrous oxide from fertilizer use).

Land Use

Climate change is attributed to land use for two main reasons. While 66% of anthropogenic CO_2 emissions over the last 250 years have resulted from burning fossil fuels, 33% have resulted from changes in land use, primarily deforestation. Deforestation both reduces the amount of carbon dioxide absorbed by deforested regions and releases greenhouse gases directly, together with aerosols, through biomass burning that frequently accompanies it. A second reason that climate change has been attributed to land use is that the terrestrial albedo is often altered by use, which leads to radiative forcing. This effect is more significant locally than globally.

Livestock and Land Use

Worldwide, livestock production occupies 70% of all land used for agriculture, or 30% of the ice-free land surface of the Earth. More than 18% of anthropogenic greenhouse gas emissions are attributed to livestock and livestock-related activities such as deforestation and increasingly fuel-intensive farming practices. Specific attributions to the livestock sector include:

- 9% of global anthropogenic carbon dioxide emissions
- 35–40% of global anthropogenic methane emissions (chiefly due to enteric fermentation and manure)
- 64% of global anthropogenic nitrous oxide emissions, chiefly due to fertilizer use.

Aerosols

With virtual certainty, scientific consensus has attributed various forms of climate change, chiefly cooling effects, to aerosols, which are small particles or droplets suspended in the atmosphere. Key sources to which anthropogenic aerosols are attributed include:

- biomass burning such as slash and burn deforestation. Aerosols produced are primarily black carbon.
- industrial air pollution, which produces soot and airborne sulfates, nitrates, and ammonium
- dust produced by land use effects such as desertification

ATTRIBUTION OF 20TH CENTURY CLIMATE CHANGE

Over the past 150 years human activities have released increasing quantities of greenhouse gases into the atmosphere. This has led to increases in mean global temperature, or global warming. Other human effects are relevant—for example, sulphate aerosols are believed to lead to cooling—and natural factors also contribute. According to the historical temperature record of the last century, the Earth's near-surface air temperature has risen around 0.74 ± 0.18 °Celsius (1.3 ± 0.32 °Fahrenheit).

A historically important question in climate change research has regarded the relative importance of human activity and non-anthropogenic causes during the period of instrumental record. In the 1995 Second Assessment Report (SAR), the IPCC made the widely quoted statement that "The balance of evidence suggests a discernible human influence on global climate". The phrase "balance of evidence" suggested the (English) common-law standard of proof required in civil as opposed to criminal courts: not as high as "beyond reasonable doubt". In 2001 the Third Assessment Report (TAR) refined this, saying "There is new and stronger evidence that most of the warming observed over the last 50 years is attributable to human activities". The 2007 fourth assessment report (WG1 AR4) strengthened this finding:

- "Anthropogenic warming of the climate system is widespread and can be detected in temperature observations taken at the surface, in the free atmosphere and in the oceans. Evidence of the effect of external influences, both anthropogenic and natural, on the climate system has continued to accumulate since the TAR."

  Over the past five decades there has been a global warming of approximately 0.65 °C (1.17 °F) at the Earth's surface. Among the possible factors that could produce changes in global mean temperature are internal variability of the climate system, external forcing, an increase in concentration of greenhouse gases, or any combination of these. Current studies indicate that the increase in greenhouse gases, most notably CO_2, is mostly responsible for the observed warming. Evidence for this conclusion includes:
- Estimates of internal variability from climate models, and reconstructions of past temperatures, indicate that the warming is unlikely to be entirely natural.
- Climate models forced by natural factors *and* increased greenhouse gases and aerosols reproduce the observed global temperature changes; those forced by natural factors alone do not.
- "Fingerprint" methods indicate that the pattern of change is closer to that expected from greenhouse gas-forced change than from natural change.
- The plateau in warming from the 1940s to 1960s can be attributed largely to sulphate aerosol cooling.

Detection vs. Attribution

Detection and attribution of climate signals, as well as its common-sense meaning, has a more precise definition within the climate change literature, as expressed by the IPCC.

Detection of a signal requires demonstrating that an observed change is statistically significantly different from that which can be explained by natural internal variability.

Attribution requires demonstrating that a signal is:

- unlikely to be due entirely to internal variability;
- consistent with the estimated responses to the given combination of anthropogenic and natural forcing
- not consistent with alternative, physically plausible explanations of recent climate change that exclude important elements of the given combination of forcings.

Detection does not imply attribution, and is easier to show than attribution. Unequivocal attribution would require controlled experiments with multiple copies of the climate system, which is not possible. Therefore, attribution, as described above, can only be done within some margin of error. For example, the IPCC's Fourth Assessment Report says "it is *extremely likely* that human activities have exerted a substantial net warming influence on climate since 1750," where "extremely likely" indicates a probability greater than 95%.

Following the publication of the Third Assessment Report (TAR) in 2001, "detection and attribution" of climate change has remained an active area of research. Some important results include:

- A review of detection and attribution studies by the International Ad Hoc Detection and Attribution Group found that "natural drivers such as solar variability and volcanic activity are at most partially responsible for the large-scale temperature changes observed over the past century, and that a large fraction of the warming over the last 50 yr can be attributed to greenhouse gas increases. Thus, the recent research supports and strengthens the IPCC Third Assessment Report conclusion that 'most of the global warming over the past 50 years is likely due to the increase in greenhouse gases.'"
- Multiple independent reconstructions of the temperature record of the past 1000 years confirm that the late 20th century is probably the warmest period in that time

- Two papers in the journal *Science* in August 2005 resolve the problem, evident at the time of the TAR, of tropospheric temperature trends. The UAH version of the record contained errors, and there is evidence of spurious cooling trends in the radiosonde record, particularly in the tropics.
- Barnett and colleagues say that the observed warming of the oceans "cannot be explained by natural internal climate variability or solar and volcanic forcing, but is well simulated by two anthropogenically forced climate models," concluding that "it is of human origin, a conclusion robust to observational sampling and model differences"

Scientific Literature and Opinion

- The attribution of climate change is discussed extensively, with references to peer-reviewed research. The Meaning of Detection and Attribution, Quantitative Comparison of Observed and Modelled Climate Change, Pattern Correlation Methods and Optimal Fingerprint Methods.
- An essay in *Science* surveyed 928 abstracts related to climate change, and concluded that most journal reports accepted the consensus.
- A 2002 paper in the *Journal of Geophysical Research* says "Our analysis suggests that the early twentieth century warming can best be explained by a combination of warming due to increases in greenhouse gases and natural forcing, some cooling due to other anthropogenic forcings, and a substantial, but not implausible, contribution from internal variability. In the second half of the century we find that the warming is largely caused by changes in greenhouse gases, with changes in sulphates and, perhaps, volcanic aerosol offsetting approximately one third of the warming."
- In 1996, in a paper in *Nature* titled "A search for human influences on the thermal structure of the atmosphere", Benjamin D. Santer et al. wrote: "The observed spatial patterns of temperature change in the free atmosphere from 1963 to 1987 are similar to those predicted by

state-of-the-art climate models incorporating various combinations of changes in carbon dioxide, anthropogenic sulphate aerosol and stratospheric ozone concentrations. The degree of pattern similarity between models and observations increases through this period. It is likely that this trend is partially due to human activities, although many uncertainties remain, particularly relating to estimates of natural variability."

- Some scientists noted for their somewhat skeptical view of global warming accept that recent climate change is mostly anthropogenic. John Christy has said that he supports the American Geophysical Union (AGU) declaration, and is convinced that human activities are the major cause of the global warming that has been measured.

FINDINGS THAT COMPLICATE ATTRIBUTION TO CO_2

Warming Sometimes Leads CO_2 Increases

Factors other than increased CO_2 concentrations can initiate warming or cooling episodes. The ice core record shows that on some occasions temperature starts rising hundreds of years before CO_2 increases. Such results confirm that the relationship between CO_2 and climate can go in both directions: changes in CO_2 concentrations affect climate, while changes in climate can affect CO_2 concentrations. One proposed mechanism for this effect is increased release of sequestered CO_2 from oceans as circulation patterns shift, perhaps abruptly, in response to climate change.

A more speculative and polemical inference sometimes drawn is that the causal relationship between temperature rises and global CO_2 concentrations is only one-way, so that historical increases in CO_2 have been nothing more than the product of independently rising temperatures. However, a strictly "one-way" view of the relationship between CO_2 and temperature contradicts basic results in physics, specifically the fact that the absorption and emission of infrared radiation by CO_2 increases as its atmospheric concentration increases.

First principles as well as empirical observation suggest that positive feedbacks from CO_2 concentrations amplify warming initially caused by other factors: Close analysis of the relationship between the two curves [i.e., temperature and CO_2] shows that, within the uncertainties of matching their timescales, the temperature led by a few centuries. This is expected, since it was changes in the Earth's orbital parameters (including the shape of its orbit around the Sun, and the tilt of Earth's axis) that caused the small initial temperature rise. This then raised atmospheric CO_2 levels, in part by outgassing from the oceans, causing the temperature to rise further. By amplifying each other's response, this "positive feedback" can turn a small initial perturbation into a large climate change. There is therefore no surprise that the temperature and CO_2 rose in parallel, with the temperature initially in advance. In the current case, the situation is different, because human actions are raising the CO_2 level, and we are starting to observe the temperature response.

Present CO_2 levels greatly exceed the range found in the ice core data. Isotopic analysis of atmospheric CO_2 confirms that fossil fuel burning is the source of most of the CO_2 increase, unlike during prior interglacial periods. As noted above, models that include increased CO_2 levels when simulating recent climate match the observed data far better than those that do not.

Warming on other Planets?

Over the last two decades, proxy evidence of local or planetary warming has been observed on Mars, Pluto, Jupiter, and Neptune's largest moon Triton. It has sometimes been asserted in the popular press that this points to a solar explanation for the recent warming on Earth. Physicist Khabibullo Abdusamatov claims that solar variation has caused global warming on Earth, and that the coincident warmings "can only be a straightline consequence of the effect of the one same factor: a long-time change in solar irradiance." This view is not accepted by other scientists. Planetary physicist Colin Wilson responded, "His views are completely at odds with the mainstream scientific opinion," and climate scientist Amato Evan stated, "the idea just isn't supported by the theory or by the observations." Charles Long of Pacific Northwest National Laboratory, who studies radiative transfer, says "That's nuts ... It doesn't make physical sense that that's the case." Jay

Pasachoff, an astronomy professor at Williams College, said that Pluto's global warming was "likely not connected with that of the Earth. The major way they could be connected is if the warming was caused by a large increase in sunlight. But the solar constant—the amount of sunlight received each second—is carefully monitored by spacecraft, and we know the Sun's output is much too steady to be changing the temperature of Pluto." Instead, scientific opinion is that these changes are caused by other factors, such as orbital irregularities or (in the case of Mars) changes in albedo as a result of dust storms.

CLIMATE MODEL

Climate models use quantitative methods to simulate the interactions of the atmosphere, oceans, land surface, and ice. They are used for a variety of purposes from study of the dynamics of the climate system to projections of future climate.

All climate models take account of incoming energy as short wave electromagnetic radiation, chiefly visible and short-wave (near) infrared, as well as outgoing energy as long wave (far) infrared electromagnetic radiation from the earth. Any imbalance results in a change in temperature.

The most talked-about models of recent years have been those relating temperature to emissions of carbon dioxide. These models project an upward trend in the surface temperature record, as well as a more rapid increase in temperature at higher altitudes.

Models can range from relatively simple to quite complex:

- A simple radiant heat transfer model that treats the earth as a single point and averages outgoing energy
- this can be expanded vertically (radiative-convective models), or horizontally
- finally, (coupled) atmosphere–ocean–sea ice global climate models discretise and solve the full equations for mass and energy transfer and radiant exchange.

This is not a full list; for example "box models" can be written to treat flows across and within ocean basins. Furthermore, other types of modelling can be interlinked, such as land use, allowing researchers to predict the interaction between climate and ecosystems.

Box Models

Box models are simplified versions of complex systems, reducing them to boxes (or reservoirs) linked by fluxes. The boxes are assumed to be mixed homogeneously. Within a given box, the concentration of any chemical species is therefore uniform. However, the abundance of a species within a given box may vary as a function of time due to the input to (or loss from) the box or due to the production, consumption or decay of this species within the box.

Simple box models, i.e. box model with a small number of boxes whose properties (e.g. their volume) do not change with time, are often useful to derive analytical formulas describing the dynamics and steady-state abundance of a species. More complex box models are usually solved using numerical techniques.

Box models are used extensively to model environmental systems or ecosystems and in studies of ocean circulation and the carbon cycle.

Zero-dimensional Models

A very simple model of the radiative equilibrium of the Earth is:

$$(1 - a)S\pi r^2 = 4\pi r^2 \varepsilon\sigma T^4$$

where

- the left hand side represents the incoming energy from the Sun
- the right hand side represents the outgoing energy from the Earth, calculated from the Stefan-Boltzmann law assuming a constant radiative temperature, T, that is to be found,

and

- S is the solar constant-the incoming solar radiation per unit area—about 1367 $W \cdot m^{-2}$
- a is the Earth's average albedo, measured to be 0.3.
- r is Earth's radius—approximately 6.371×10^6m
- π is the mathematical constant (3.141...)
- σ is the Stefan-Boltzmann constant—approximately 5.67×10^{-8} $J \cdot K^{-4} \cdot m^{-2} \cdot s^{-1}$

- ε is the effective emissivity of earth, about 0.612

The constant πr^2 can be factored out, giving

$(1 - a)S = 4\varepsilon\sigma T^4$

Solving for the temperature,

$$T = \sqrt[4]{\frac{(1-a)S}{4\epsilon\sigma}}$$

This yields an average earth temperature of 288 K (15 °C; 59 °F). This is because the above equation represents the effective *radiative* temperature of the Earth (including the clouds and atmosphere). The use of effective emissivity and albedo account for the greenhouse effect.

This very simple model is quite instructive, and the only model that could fit on a page. For example, it easily determines the effect on average earth temperature of changes in solar constant or change of albedo or effective earth emissivity. Using the simple formula, the percent change of the average amount of each parameter, considered independently, to cause a one degree Celsius change in steady-state average earth temperature is as follows:

- Solar constant 1.4%
- Albedo 3.3%
- Effective emissivity 1.4%.

The average emissivity of the earth is readily estimated from available data. The emissivities of terrestrial surfaces are all in the range of 0.96 to 0.99 (except for some small desert areas which may be as low as 0.7). Clouds, however, which cover about half of the earth's surface, have an average emissivity of about 0.5 (which must be reduced by the fourth power of the ratio of cloud absolute temperature to average earth absolute temperature) and an average cloud temperature of about 258 K ("15 °C; 5 °F). Taking all this properly into account results in an effective earth emissivity of about 0.64 (earth average temperature 285 K (12 °C; 53 °F)).

This simple model readily determines the effect of changes in solar output or change of earth albedo or effective earth emissivity on average earth temperature. It says nothing, however about what might cause these things to change. Zero-dimensional models do not address the temperature distribution on the earth or the factors that move energy about the earth.

Radiative-Convective Models

The zero-dimensional model above, using the solar constant and given average earth temperature, determines the effective earth emissivity of long wave radiation emitted to space. This can be refined in the vertical to a zero-dimensional radiative-convective model, which considers two processes of energy transport:

- upwelling and downwelling radiative transfer through atmospheric layers that both absorb and emit infrared radiation
- upward transport of heat by convection (especially important in the lower troposphere).

The radiative-convective models have advantages over the simple model: they can determine the effects of varying greenhouse gas concentrations on effective emissivity and therefore the surface temperature. But added parameters are needed to determine local emissivity and albedo and address the factors that move energy about the earth.

Higher Dimension Models

The zero-dimensional model may be expanded to consider the energy transported horizontally in the atmosphere. This kind of model may well be zonally averaged. This model has the advantage of allowing a rational dependence of local albedo and emissivity on temperature-the poles can be allowed to be icy and the equator warm-but the lack of true dynamics means that horizontal transports have to be specified.

EMICs (Earth-system Models of Intermediate Complexity)

Depending on the nature of questions asked and the pertinent time scales, there are, on the one extreme, conceptual, more inductive models, and, on the other extreme, general circulation models operating at the highest spatial and temporal resolution currently feasible. Models of intermediate complexity bridge the gap. One example is the Climber-3 model. Its atmosphere is a 2.5-dimensional statistical-dynamical model with 7.5° × 22.5° resolution and time step of 1/2 a day; the ocean is MOM-3 (Modular Ocean Model) with a 3.75° × 3.75° grid and 24 vertical levels.

GCMs (Global Climate Models or General circulation models)

Three (or more properly, four since time is also considered) dimensional GCM's discretise the equations for fluid motion and energy transfer and integrate these over time. They also contain parametrisations for processes—such as convection—that occur on scales too small to be resolved directly.

Atmospheric GCMs (AGCMs) model the atmosphere and impose sea surface temperatures as boundary conditions. Coupled atmosphere-ocean GCMs (AOGCMs, e.g. HadCM3, EdGCM, GFDL CM2.X, ARPEGE-Climat) combine the two models. The first general circulation climate model that combined both oceanic and atmospheric processes was developed in the late 1960s at the NOAA Geophysical Fluid Dynamics Laboratory AOGCMs represent the pinnacle of complexity in climate models and internalise as many processes as possible. However, they are still under development and uncertainties remain. They may be coupled to models of other processes, such as the carbon cycle, so as to better model feedback effects.

Most recent simulations show "plausible" agreement with the measured temperature anomalies over the past 150 years when forced by natural forcings alone, but better agreement is achieved when observed changes in greenhouse gases and aerosols are also included.

Research and Development

There are three major types of institution where climate models are developed, implemented and used:

- National meteorological services. Most national weather services have a climatology section.
- Universities. Relevant departments include atmospheric sciences, meteorology, climatology, and geography.
- National and international research laboratories. Examples include the National Centre for Atmospheric Research (NCAR, in Boulder, Colorado, USA), the

Geophysical Fluid Dynamics Laboratory (GFDL, in Princeton, New Jersey, USA), the Hadley Centre for Climate Prediction and Research (in Exeter, UK), the Max Planck Institute for Meteorology in Hamburg, Germany, or the Laboratoire des Sciences du Climat et de l'Environnement (LSCE), France, to name but a few.

The World Climate Research Programme (WCRP), hosted by the World Meteorological Organization (WMO), coordinates research activities on climate modelling worldwide.

CLIMATE PREDICTION CENTRE

The Climate Prediction Centre (CPC) is one of the National Centers for Environmental Prediction, which are a part of NOAA's National Weather Service. It is located in Camp Springs, Maryland.

Products

The CPC's products are operational predictions of climate variability, real-time monitoring of global climate, and attribution of the origins of major climate anomalies. The products cover time scales from a week to seasons, and cover the land, the ocean, and the atmosphere, extending into the stratosphere.

These climate services are available for users in government, the public and private industry, both in this country and abroad. Applications include the mitigation of weather related natural disasters and uses for social and economic good in agriculture, energy, transportation, water resources, and health. Continual product improvements are supported through diagnostic research, increasing use of models, and interactions with user groups. Some specific products include:

- 3-Month Temperature and Precipitation
  - * Outlooks
  - * Discussions
- 1-Month Temperature and Precipitation
  - * Outlooks
  - * Discussions
- 6 to 10-Day and 8 to 14-Day Products
  - * Temperature and Precipitation Anomaly

  * Excessive Heat Outlook
  * Maximum Heat Index Prediction
- 3-Month Probability of Exceedance
  * Temperature
  * Precipitation
  * Heating and Cooling Degree Days
- Hurricane Season Outlook
  * Atlantic basin
  * Pacific basin
- U.S. Drought
  * Outlook
  * Discussion
- International Support
  * Weekly Afghan Hazards
  * Weekly Africa Hazards
  * Weekly Central America Hazards
  * Weekly Haiti Hazards.

History

The roots of modern climate prediction can be traced to the late 18th century. One of the nation's first applied climatologists was Thomas Jefferson, the third President of the United States. A century later, the federal government assigned to the Army Signal Corps the mission to define the climate of the regions of the country being opened for farming. In 1890, the United States Department of Agriculture (USDA) created the Weather Bureau climate and crops services which began publishing the Weather and Crops Weekly Bulletin, which the CPC in conjunction with the USDA still publishes today.

In 1970, various federal weather and climate functions were consolidated into the National Weather Service (NWS) and placed in a new agency called the National Oceanic and Atmospheric Administration (NOAA). In the 1980s the National Weather Service established the Climate Prediction Centre, known at the time as the Climate Analysis Centre (CAC). The CPC is best known for its United States climate forecasts based on El Niño and La Niña conditions in the tropical Pacific.

TROPICAL MARINE CLIMATE

Islands and coastal areas 10° to 20° north or south of the equator usually have a tropical marine climate. The climate of the tropical marine system is influenced by the sea. There are two main seasons the wet season and the dry season. The annual rainfall is 1000 to over 1500 mm. The temperature ranges from 25°C to 35°C. The trade winds blow all year round. The trade winds are moist as they have passed over warm seas.

Wet Season

The wet season of the tropical marine climate occurs during the period when the conditions of the atmosphere are not stable. At this time, the regions (10° to 20° north or south of the equator) experience tropical disturbances. Around this time islands like Jamaica and Grenada are affected by the ITCZ (Intertropical Convergence Zone). The most rainfall comes between July to September.

Dry Season

The dry season occurs when the conditions in the atmosphere are stable. During the dry season there is less rainfall than in the wet season. Around this time the tropical marine regions is influenced by anti cyclones. There is little difference between the wet and dry seasons.

Adaptations

The ecosystems of the tropical marine climate have to adapt to the dry season. Plants during the dry season must conserve water/moisture. However the extent of the adaptation depends much on the annual rainfall. Hygrophytic ecosystems occur when there is a short dry period with a few rain showers. The soil in this ecosystem holds adequate water for plant growth. Most of the tropical marine ecosystems are close to true rain forests.

Mesophytic Ecosystem

The mesophytic ecosystem is also known as a semi-evergreen forest. It is found where there is a long dry season that has little rainfall. There is less vegetation than in a rainforest and the layer structure is simpler. There are only 2 tree stories, trees shed their leaves or have very small leaves.

This provides the plants a way to conserve moisture. There are fewer epiphytes than a rain forest has as the canopy is dry. In the dry season the ground is covered by leaves that will not decay until the soil is moist. The trees often flower during the dry season and start to grow during the wet season. The soil is usually latasol.

Xerophytic Ecosystem

The xerophytic ecosystem is also known as dry woodland. It is found in areas of rain shadow in the tropical marine climate. This ecosystem often develops soils that drain quickly. The dry woodland is very different than the rainforest. The biomass is a lot less than a rainforest as there is little rain. The tallest of trees are only 15 to 25 meters high in the dry woodland. Dry woodland trees either have small leaves or shed their leaves. The trees have very thick bark and the trunks are crooked.

Variations

Mangroves grow in coastal wet lands which are called hydrophytic ecosystems. The vegetation at the coast arc usually adapted to sandy soil. The montane forests and elfin wood lands grow on the cool, moist, mountainous regions.

9

Outer Space

Outer space (often simply called space) is the void that exists beyond any celestial body including the Earth. It is not completely empty (i.e. a perfect vacuum), but contains a low density of particles, predominantly hydrogen plasma, as well as electromagnetic radiation, magnetic fields, and neutrinos. Theoretically, it also contains dark matter and dark energy.

Discovery

In 350 BC, Greek philosopher Aristotle suggested that *nature abhors a vacuum*, a principle that became known as the *horror vacui*. Based on this idea that a vacuum could not exist, it was widely held for many centuries that space could not be empty. As late as the seventeenth century, the French philosopher René Descartes argued that the entirety of space must be filled. It became known to Galileo Galilei that air had weight and so was subject to gravity. He also demonstrated that there was an established force that resisted the formation of a vacuum. However, it would remain for his pupil Evangelista Torricelli to create an apparatus that would produce a vacuum. At the time this experiment created a scientific sensation in Europe. The French mathematician Blaise Pascal reasoned that if the column of mercury was suspended by air then the column ought to be shorter at higher altitude. His brother in law, Florin Périer, repeated the experiment on the Puy-de-Dôme mountain in central France and found that the column was shorter by three inches. This decrease in pressure was further demonstrated by carrying a half-full balloon up a mountain and watching it gradually inflate, then deflate upon descent. These and other experiments were used to overthrow the principle of *horror vacui*.

Further work on the physics of the vacuum was performed by Otto von Guericke. He correctly noted that the atmosphere of the Earth surrounds the planet like a shell, with the density gradually declining with altitude. He concluded that there must be a vacuum between the Earth and the Moon.

Early speculations as to the infinite dimension of space was performed in the sixteenth century by the Italian philosopher Giordano Bruno. He extended the Copernican heliocentric cosmology to the concept of an infinite universe that is filled with a substance he called aether, which did not cause resistance to the motions of heavenly bodies. English philosopher William Gilbert arrived at a similar conclusion, arguing that the stars are visible to us only because they are surrounded by a thin aether or a void. This concept of an aether originated with ancient Greek philosophers, including Aristotle, who conceived of it as the medium through which the heavenly bodies moved.

The concept of a universe filled with a luminiferous aether remained in vogue among some scientists up until the twentieth century. This form of aether was viewed as the medium through which light could propagate. In 1887, the Michelson-Morley experiment was carried out as an attempt to detect the Earth's motion through this medium by looking for changes in the speed of light based on the direction of the planet's motion. However, the null result indicated something was wrong with the concept. Since then the idea of the luminiferous aether had essentially been abandoned, to be replaced by Albert Einstein's theory of special relativity. The latter held that the speed of light is a constant in a vacuum, regardless of the observer's motion or frame of reference.

The first professional astronomer to support the concept of an infinite universe was the Englishman Thomas Digges in 1576. However, the true scale of the universe remained unknown until the first successful measurement of the distance to a nearby star was performed in 1838 by the German astronomer Friedrich Bessel. He showed that the star 61 Cygni had a parallax of just 0.31 arcseconds (compared to the modern value of 0.2873). This corresponded to a distance of over 10 light years. The distance scale to the Andromeda galaxy was determined in 1923 by American astronomer Edwin Hubble when he measured the brightness of cepheid variables within that galaxy. This established that the Andromeda galaxy, and by extension all galaxies, lay well outside the Milky Way.

The modern concept of outer space is based upon the Big Bang cosmology, which was first proposed in 1931 by the Belgian physicist Georges Lemaître. This theory holds that the observable universe originated from a very compact form that has since undergone continuous expansion. Matter that remained following the initial expansion has since undergone gravitational collapse to create stars, galaxies and other astronomical objects, leaving behind a deep vacuum that forms what is now called outer space.

The term *outer space* was first recorded by the English poet Lady Emmeline Stuart-Wortley in her poem "The Maiden of Moscow" in 1842, and later popularized in the writings of HG Wells in 1901. The shorter term *space* is actually older, first used to mean the region beyond Earth's sky in John Milton's *Paradise Lost* in 1667.

Environment

Outer space is the closest natural approximation of a perfect vacuum. It has effectively no friction, allowing stars, planets and moons to move freely along ideal gravitational trajectories. However, even in the deep vacuum of intergalactic space there are still a few hydrogen atoms per cubic meter. By comparison, the air we breathe contains about 10^{25} molecules per cubic meter. The sparse density of matter in outer space means that electromagnetic radiation can travel great distances without being scattered; the mean free path for a photon in intergalactic space is about 10^{23} km, or 10 billion light years. The deep vacuum of space could make it an attractive environment for certain industrial processes, for instance those that require ultraclean surfaces. Stars, planets, asteroids and moons keep their atmospheres by gravitational attraction, and as such, atmospheres have no clearly delineated boundary: the density of atmospheric gas simply decreases with distance from the object. The Earth's atmospheric pressure drops to about 1 Pa at 100 kilometres (62 mi) of altitude. This is known as the

Kármán line, a common definition of the boundary with outer space. Beyond this line, isotropic gas pressure rapidly becomes insignificant when compared to radiation pressure from the Sun and the dynamic pressure of the solar wind, so the definition of pressure becomes difficult to interpret. The thermosphere in this range has large gradients of pressure, temperature and composition, and varies greatly due to space weather. Astrophysicists prefer to use number density to describe these environments, in units of particles per cubic centimetre.

Temperature

All of the observable Universe is filled with photons that were created during the Big Bang, which is known as the cosmic microwave background radiation (CMB). There is quite likely a correspondingly large number of neutrinos called the cosmic neutrino background. The current black body temperature of this photon radiation is about 3 K ("270 °C; "454 °F). Some regions of outer space can contain highly energetic particles that have a much higher temperature than the CMB.

Effect on Human Bodies

Contrary to popular belief, a person suddenly exposed to the vacuum would not explode, freeze to death or die from boiling blood. Air would immediately leave the lungs due to the enormous pressure gradient. Any oxygen dissolved in the blood would empty into the lungs to try to equalize the partial pressure gradient. Once the deoxygenated blood arrived at the brain, death would quickly follow.

Humans and animals exposed to vacuum will lose consciousness after a few seconds and die of hypoxia within minutes. Blood and other body fluids do boil when their pressure drops below 6.3 kPa, the vapour pressure of water at body temperature. This condition is called ebullism. The steam may bloat the body to twice its normal size and slow circulation, but tissues are elastic and porous enough to prevent rupture. Ebullism is slowed by the pressure containment of blood vessels, so some blood remains liquid. Swelling and ebullism can be reduced by containment in a flight suit. Shuttle astronauts wear a fitted elastic garment called the Crew Altitude Protection Suit (CAPS) which prevents ebullism at pressures as low as

2 kPa. Water vapour would also rapidly evaporate off from exposed areas such as the lungs, cornea of the eye and mouth, cooling the body. Rapid evaporative cooling of the skin will create frost, particularly in the mouth, but this is not a significant hazard. Space may be cold, but it's mostly vacuum and transfers heat ineffectively; as a result the main temperature regulation concern for space suits is how to get rid of naturally generated body heat.

Cold or oxygen-rich atmospheres can sustain life at pressures much lower than atmospheric, as long as the density of oxygen is similar to that of standard sea-level atmosphere. The colder air temperatures found at altitudes of up to 3 kilometres (1.9 mi) generally compensate for the lower pressures there. Above this altitude, oxygen enrichment is necessary to prevent altitude sickness, and spacesuits are necessary to prevent ebullism above 19 kilometres (12 mi). Most spacesuits use only 20 kPa of pure oxygen, just enough to sustain full consciousness. This pressure is high enough to prevent ebullism, but simple evaporation of blood can still cause decompression sickness and gas embolisms if not managed.

Rapid decompression can be much more dangerous than vacuum exposure itself. Even if the victim does not hold his breath, venting through the windpipe may be too slow to prevent the fatal rupture of the delicate alveoli of the lungs. Eardrums and sinuses may be ruptured by rapid decompression, soft tissues may bruise and seep blood, and the stress of shock will accelerate oxygen consumption leading to hypoxia. Injuries caused by rapid decompression are called barotrauma. A pressure drop as small as 13 kPa, which produces no symptoms if it is gradual, may be fatal if it occurs suddenly.

Boundary

There is no clear boundary between Earth's atmosphere and space, as the density of the atmosphere gradually decreases as the altitude increases. There are several designated scientific boundaries, namely:

- The Fédération Aéronautique Internationale has established the Kármán line at an altitude of 100 kilometres (62 mi) as a working definition for the boundary between aeronautics and astronautics. This is used because above an altitude of roughly 100 km, as Theodore von Kármán calculated, a vehicle would have to travel faster than orbital velocity in order to derive sufficient aerodynamic lift from the atmosphere to support itself.

- The United States designates people who travel above an altitude of 50 miles (80 km) as astronauts.
- NASA's mission control uses 76 miles (122 km) as their re-entry altitude, which roughly marks the boundary where atmospheric drag becomes noticeable (depending on the ballistic coefficient of the vehicle), thus leading shuttles to switch from steering with thrusters to maneuvering with air surfaces.

In 2009, scientists at the University of Calgary reported detailed measurements with an instrument called the Supra-Thermal Ion Imager (an instrument that measures the direction and speed of ions), which allowed them to determine that space begins 118 kilometres (73 mi) above Earth. The boundary represents the midpoint of a gradual transition over tens of kilometres from the relatively gentle winds of the Earth's atmosphere to the more violent flows of charged particles in space, which can reach speeds well over 600 miles per hour (1,000 km/h). This was only the second time that direct measurements of charged particle flows have been conducted at this region, which is too high for balloons and too low for satellites. It was however the first study to include all the relevant elements for this kind of determination – for example, the upper atmospheric winds.

The instrument was carried by the JOULE-II rocket on January 19, 2007, and traveled to an altitude of about 124 miles (200 km). From there it collected data while it was moving through the "edge of space".

Legal Status

The Outer Space Treaty provides the basic framework for international space law. This treaty covers the legal use of outer space by nation states, and includes in its definition of *outer space* the Moon and other celestial bodies. The treaty states that outer space is free for all nation states to explore and is not subject to claims of national sovereignty. It also prohibits the deployment of nuclear weapons in outer space. The treaty was passed by the United Nations General Assembly in 1963 and signed in 1967 by the USSR, the United States of America and the United Kingdom. As of January 1, 2008 the treaty has been ratified by 98 states and signed by an additional 27 states.

Between 1958 and 2008, outer space has been the subject of multiple resolutions by the United Nations General Assembly. Of these, more than 50 have been concerning the international co-operation in the peaceful uses of outer space and preventing an arms race in space. Four additional space law treaties have been negotiated and drafted by the UN's Committee on the Peaceful Uses of Outer Space. The 1979 Moon Treaty turned the jurisdiction of all heavenly bodies (including the orbits around such bodies) over to the international community. However, this treaty has not been ratified by any nation that currently practices manned spaceflight.

Space versus Orbit

To perform an orbit, a spacecraft must travel faster than a sub-orbital spaceflight. A spacecraft has not entered orbit until it is traveling with a sufficiently great horizontal velocity such that the acceleration due to gravity on the spacecraft is less than or equal to the centripetal acceleration being caused by its horizontal velocity. So to enter orbit, a spacecraft must not only reach space, but must also achieve a sufficient orbital speed (angular velocity). For a low-Earth orbit, this is about 7,900 m/s (28,400 km/h; 17,700 mph); by contrast, the fastest airplane speed ever achieved (excluding speeds achieved by deorbiting spacecraft) was 2,200 m/s (7,900 km/h; 4,900 mph) in 1967 by the North American X-15.

Konstantin Tsiolkovsky was the first person to realize that, given the energy available from any available chemical fuel, a several-stage rocket would be required. The escape velocity to pull free of Earth's gravitational field altogether and move into interplanetary space is about 11,000 m/s (39,600 km/h; 24,600 mph) The energy required to reach velocity for low Earth orbit (32 MJ/kg) is about twenty times the energy required simply to climb to the corresponding altitude (10 kJ/(km·kg)).

There is a major difference between sub-orbital and orbital spaceflights. The minimum altitude for a stable orbit around Earth (that is, one without significant atmospheric drag) begins at around 350 kilometres (220 mi) above mean sea level. A common misunderstanding about the boundary to space is that orbit occurs simply by reaching this altitude. Achieving orbital speed can theoretically occur at any altitude, although atmospheric drag precludes an orbit that is too low. At sufficient speed, an airplane would need a way to keep it from flying off into space, but at present, this speed is several times greater than anything within reasonable technology.

A common misconception is that people in orbit are outside Earth's gravity because they are "floating". They are floating because they are in "free fall": they are accelerating toward Earth, along with their spacecraft, but are simultaneously moving sideways fast enough that the "fall" away from a straight-line path merely keeps them in orbit at a constant distance above Earth's surface. Earth's gravity reaches out far past the Van Allen belt and keeps the Moon in orbit at an average distance of 384,403 kilometres (238,857 mi).

Regions

Space is not a perfect vacuum: its different regions are defined by the various atmospheres and "winds" that dominate within them, and extend to the point at which those winds give way to those beyond. Geospace extends from Earth's atmosphere to the outer reaches of Earth's magnetic field, whereupon it gives way to the solar wind of interplanetary space. Interplanetary space extends to the heliopause, whereupon the solar wind gives way to the winds of the interstellar medium. Interstellar space then continues to the edges of the galaxy, where it fades into the intergalactic void.

Geospace

Geospace is the region of outer space near the Earth. Geospace includes the upper region of the atmosphere, as well as the ionosphere and magnetosphere. The Van Allen radiation belts also lie within the geospace. The region between Earth's atmosphere and the Moon is sometimes referred to as cis-lunar space. Although it meets the definition of outer space, the atmospheric density within the first few hundred kilometres above the Kármán line is still sufficient to produce significant drag on satellites. Most artificial satellites operate in this region called low earth orbit and must fire their engines every few days to maintain orbit. The drag here is low enough that it could theoretically be overcome by radiation pressure on solar sails, a proposed propulsion system for interplanetary travel.

Geospace is populated by electrically charged particles at very low densities, the motions of which are controlled by the Earth's magnetic field. These plasmas form a medium from which storm-like disturbances powered by the solar wind can drive electrical currents into the Earth's upper atmosphere. During geomagnetic storms two regions of geospace, the radiation belts and the ionosphere, can become strongly disturbed. These storms increase fluxes of energetic electrons that can permanently damage satellite electronics, disrupting telecommunications and GPS technologies, and can also be a hazard to astronauts, even in low-Earth orbit. They also create aurorae seen near the magnetic poles.

Geospace contains material left over from previous manned and unmanned launches that are a potential hazard to spacecraft. Some of this debris re-enters Earth's atmosphere periodically. The absence of air makes geospace (and the surface of the Moon) ideal locations for astronomy at all wavelengths of the electromagnetic spectrum, as evidenced by the spectacular pictures sent back by the Hubble Space Telescope, allowing light from about 13.7 billion years ago — almost to the time of the Big Bang — to be observed.

The outer boundary of geospace is the interface between the magnetosphere and the solar wind. The inner boundary is the ionosphere. Alternately, geospace is the region of space between the Earth's upper atmosphere and the outermost reaches of the Earth's magnetic field.

Interplanetary

Interplanetary space, the space around the Sun and planets of the Solar System, is the region dominated by the interplanetary medium, which extends out to the heliopause where the influence of the galactic environment starts to dominate over the magnetic field and particle flux from the Sun. Interplanetary space is defined by the solar wind, a continuous stream of charged particles emanating from the Sun that creates a very tenuous atmosphere (the heliosphere) for billions of miles into space. This wind has a particle density of 5–10 protons/cm^3 and is moving at a velocity of 350–400 km/s. The distance and strength of the heliopause varies depending on the activity level of the solar wind. The discovery since 1995 of extrasolar planets means that other stars must possess their own interplanetary media.

The volume of interplanetary space is an almost pure vacuum, with a mean free path of about one astronomical unit at the orbital distance of the Earth. However, this space is not completely empty, and is sparsely filled with cosmic rays, which include ionized atomic nuclei and various subatomic particles. There is also gas, plasma and dust, small meteors, and several dozen types of organic molecules discovered to date by microwave spectroscopy. Interplanetary space contains the magnetic field generated by the Sun. There are also magnetospheres generated by planets such as Jupiter, Saturn and the Earth that have their own magnetic fields. These are shaped by the influence of the solar wind into the approximation of a teardrop shape, with the long tail extending outward behind the planet. These magnetic fields can trap particles from the solar wind and other sources, creating belts of magnetic particles such as the Van Allen Belts. Planets without magnetic fields, such as Mars and Mercury, but excluding Venus, have their atmospheres gradually eroded by the solar wind.

Interstellar

Interstellar space is the physical space within a galaxy not occupied by stars or their planetary systems. The interstellar medium resides—by definition—in interstellar space.

Intergalactic

Intergalactic space is the physical space between galaxies. Generally free of dust and debris, intergalactic space is very close to a total vacuum. The space between galaxy clusters, called the voids, is probably nearly empty. Some theories put the average density of the Universe as the equivalent of one hydrogen atom per cubic meter. The density of the universe, however, is clearly not uniform; it ranges from relatively high density in galaxies (including very high density in structures within galaxies, such as planets, stars, and black holes) to conditions in vast voids that have much lower density than the universe's average.

Surrounding and stretching between galaxies, there is a rarefied plasma that is thought to possess a cosmic filamentary structure and that is slightly denser than the average density in the universe. This material is called the *intergalactic medium (IGM)* and is mostly ionized hydrogen; i.e. a plasma consisting of equal numbers of electrons and protons. The IGM is thought to exist at a density of 10 to 100 times the average density of the universe (10 to 100 hydrogen atoms per cubic meter). It reaches densities as high as 1000 times the average density of the universe in rich clusters of galaxies.

The reason the IGM is thought to be mostly ionized gas is that its temperature is thought to be quite high by terrestrial standards (though some parts of it are only "warm" by astrophysical standards). As gas falls into the Intergalactic Medium from the voids, it heats up to temperatures of 10^5 K to 10^7 K, which is high enough for the bound electrons to escape from the hydrogen nuclei upon collisions. At these temperatures, it is called the Warm-Hot Intergalactic Medium (WHIM). Computer simulations indicate that on the order of half the atomic matter in the universe might exist in this warm-hot, rarefied state. When gas falls from the filamentary structures of the WHIM into the galaxy clusters at the intersections of the cosmic filaments, it can heat up even more, reaching temperatures of 10^8 K and above.

10

The Moon

The Moon is Earth's only natural satellite and is the fifth largest satellite in the Solar System. It is the largest natural satellite in the Solar System relative to the size of its planet, a quarter the diameter of Earth and 1/81 its mass, and is the second densest satellite after Io. It is in synchronous rotation with Earth, always showing the same face; the near side is marked with dark volcanic maria among the bright ancient crustal highlands and prominent impact craters. It is the brightest object in the sky after the Sun, although its surface is actually very dark, with a similar reflectance to coal. Its prominence in the sky and its regular cycle of phases have since ancient times made the Moon an important cultural influence on language, the calendar, art and mythology. The Moon's gravitational influence produces the ocean tides and the minute lengthening of the day. The Moon's current orbital distance, about thirty times the diameter of the Earth, causes it to be the same size in the sky as the Sun—allowing the Moon to cover the Sun precisely in total solar eclipses.

The Moon is the only celestial body on which humans have made a manned landing. While the Soviet Union's Luna programme was the first to reach the Moon with unmanned spacecraft, the United States' NASA Apollo program achieved the only manned missions to date, beginning with the first manned lunar orbiting mission by Apollo 8 in 1968, and six manned lunar landings between 1969 and 1972—the first being Apollo 11 in 1969. These missions returned over 380 kg of lunar rocks, which have been used to develop a detailed geological understanding of the Moon's origins (it is thought to have formed some 4.5 billion years ago in a giant impact event involving Earth), the formation of its internal structure, and its subsequent history.

After the Apollo 17 mission in 1972, the Moon has been visited only by unmanned spacecraft, notably by Soviet Lunokhod rovers. Since 2004, Japan, China, India, the United States, and the European Space Agency have each sent lunar orbiters. These spacecraft have contributed to confirming the discovery of lunar water ice in permanently shadowed craters at the poles and bound into the lunar regolith. Future manned missions to the Moon are planned but not yet underway; the Moon remains, under the Outer Space Treaty, free to all nations to explore for peaceful purposes.

Formation

Several mechanisms have been proposed for the Moon's formation 4.527 ± 0.010 billion years ago, some 30–50 million years after the origin of the Solar System. These include the fission of the Moon from the Earth's crust through centrifugal forces, which would require too great an initial spin of the Earth, the gravitational capture of a pre-formed Moon, which would require an unfeasibly extended atmosphere of the Earth to dissipate the energy of the passing Moon, and the co-formation of the Earth and the Moon together in the primordial accretion disk, which does not explain the depletion of metallic iron in the Moon. These hypotheses also cannot account for the high angular momentum of the Earth–Moon system.

The prevailing hypothesis today is that the Earth–Moon system formed as a result of a giant impact: a Mars-sized body hit the nearly formed proto-Earth, blasting material into orbit around the proto-Earth, which accreted to form the Moon. Giant impacts are thought to have been common in the early Solar System. Computer simulations modelling a giant impact are consistent with measurements of the angular momentum of the Earth–Moon system, and the small size of the lunar core; they also show that most of the Moon came from the impactor, not from the proto-Earth. However, meteorites show that other inner Solar System bodies such as Mars and Vesta have very different oxygen and tungsten isotopic compositions to the

Earth, while the Earth and Moon have near-identical isotopic compositions. Post-impact mixing of the vaporized material between the forming Earth and Moon could have equalized their isotopic compositions, although this is debated.

The large amount of energy released in the giant impact event and the subsequent reaccretion of material in Earth orbit would have melted the outer shell of the Earth, forming a magma ocean. The newly formed Moon would also have had its own lunar magma ocean; estimates for its depth range from about 500 km to the entire radius of the Moon.

PHYSICAL CHARACTERISTICS

Internal Structure

Chemical composition of the lunar surface regolith (derived from crustal rocks)

Compound	*Formula*	*Composition (wt %)*	
		Maria	*Highlands*
silica	SiO_2	45.4%	45.5%
alumina	Al_2O_3	14.9%	24.0%
lime	Ca	11.8%	15.9%
iron(II) oxide	Fe	14.1%	5.9%
magnesia	Mg	9.2%	7.5%
titanium dioxide	TiO_2	3.9%	0.6%
sodium oxide	Na_2	0.6%	0.6%
Total		99.9%	100.0%

The Moon is a differentiated body: it has a geochemically distinct crust, mantle, and core. This structure is thought to have developed through the fractional crystallization of a global magma ocean shortly after the Moon's formation 4.5 billion years ago. Crystallization of this magma ocean would have created a mafic mantle from the precipitation and sinking of the minerals olivine, clinopyroxene, and orthopyroxene; after about three-quarters of the magma ocean had crystallised, lower-density plagioclase minerals could form and float into a crust on top. The final liquids to crystallise would have been initially sandwiched between the crust and mantle, with a high abundance of incompatible and heat-producing elements.

Consistent with this, geochemical mapping from orbit shows the crust is mostly anorthosite, and moon rock samples of the flood lavas erupted on the surface from partial melting in the mantle confirm the mafic mantle composition, which is more iron rich than that of Earth. Geophysical techniques suggest that the crust is on average ~50 km thick.

The Moon is the second densest satellite in the Solar System after Io. However, the core of the Moon is small, with a radius of about 350 km or less; this is only ~20% the size of the Moon, in contrast to the ~50% of most other terrestrial bodies. Its composition is not well constrained, but it is probably metallic iron alloyed with a small amount of sulphur and nickel; analyses of the Moon's time-variable rotation indicate that it is at least partly molten.

Surface Geology

The Moon is in synchronous rotation: it rotates about its axis in about the same time it takes to orbit the Earth. This results in it nearly always keeping the same face turned towards the Earth. The Moon used to rotate at a faster rate, but early in its history, its rotation slowed and became locked in this orientation as a result of frictional effects associated with tidal deformations caused by the Earth. The side of the Moon that faces Earth is called the near side, and the opposite side the far side. The far side is often called the "dark side," but in fact, it is illuminated as often as the near side: once per lunar day, during the new Moon phase we observe on Earth when the near side is dark. The topography of the Moon has been measured with laser altimetry and stereo image analysis. The most visible topographic feature is the giant far side South Pole – Aitken basin, some 2,240 km in diameter, the largest crater on the Moon and the largest known crater in the Solar System. At 13 km deep, its floor is the lowest elevation on the Moon. The highest elevations are found just to its north-east, and it has been suggested that this area might have been thickened by the oblique formation impact of South Pole – Aitken. Other large impact basins, such as Imbrium, Serenitatis, Crisium, Smythii, and Orientale, also possess regionally low elevations and elevated rims. The lunar far side is on average about 1.9 km higher than the near side.

Volcanic Features

The dark and relatively featureless lunar plains which can clearly be seen with the naked eye are called *maria* (Latin for "seas"; singular *mare*), since they were believed by ancient astronomers to be filled with water. They are now known to be vast solidified pools of ancient basaltic lava. While similar to terrestrial basalts, the mare basalts have much higher abundances of iron and are completely lacking in minerals altered by water. The majority of these lavas erupted or flowed into the depressions associated with impact basins. Several geologic provinces containing shield volcanoes and volcanic domes are found within the near side maria.

Maria are found almost exclusively on the near side of the Moon, covering 31% of the surface on the near side, compared with a few scattered patches on the far side covering only 2%. This is thought to be due to a concentration of heat-producing elements under the crust on the near side, seen on geochemical maps obtained by *Lunar Prospector*'s gamma-ray spectrometer, which would have caused the underlying mantle to heat up, partially melt, rise to the surface and erupt. Most of the Moon's mare basalts erupted during the Imbrian period, 3.0–3.5 billion years ago, although some radiometrically dated samples are as old as 4.2 billion years, and the youngest eruptions, dated by crater counting, appear to have been only 1.2 billion years ago.

The lighter-coloured regions of the Moon are called *terrae*, or more commonly *highlands*, since they are higher than most maria. They have been radiometrically dated as forming 4.4 billion years ago, and may represent plagioclase cumulates of the lunar magma ocean. In contrast to the Earth, no major lunar mountains are believed to have formed as a result of tectonic events.

Impact Craters

The other major geologic process that has affected the Moon's surface is impact cratering, with craters formed when asteroids and comets collide with the lunar surface. There are estimated to be roughly 300,000 craters wider than 1 km on the Moon's near side alone. Some of these are named for scholars, scientists, artists and explorers. The lunar geologic timescale is based on the most prominent impact events, including

Nectaris, Imbrium, and Orientale, structures characterized by multiple rings of uplifted material, typically hundreds to thousands of kilometres in diameter and associated with a broad apron of ejecta deposits that form a regional stratigraphic horizon. The lack of an atmosphere, weather and recent geological processes mean that many of these craters are well-preserved. While only a few multi-ring basins have been definitively dated, they are useful for assigning relative ages. Since impact craters accumulate at a nearly constant rate, counting the number of craters per unit area can be used to estimate the age of the surface. The radiometric ages of impact-melted rocks collected during the Apollo missions cluster between 3.8 and 4.1 billion years old: this has been used to propose a Late Heavy Bombardment of impacts.

Blanketed on top of the Moon's crust is a highly comminuted (broken into ever smaller particles) and impact gardened surface layer called regolith, formed by impact processes. The finer regolith, the lunar soil of silicon dioxide glass, has a texture like snow and smell like spent gunpowder. The regolith of older surfaces is generally thicker than for younger surfaces: it varies in thickness from 10–20 m in the highlands and 3–5 m in the maria. Beneath the finely comminuted regolith layer is the *megaregolith*, a layer of highly fractured bedrock many kilometres thick.

Presence of Water

Liquid water cannot persist at the Moon's surface, and water vapour quickly evaporates, breaks up through photodissociation due to sunlight, and is lost to space. However, scientists have thought since the 1960s that water ice, deposited by impacting comets or produced by the reaction of oxygen-rich lunar rocks and hydrogen in the solar wind, could survive in the cold, permanently shadowed craters at the Moon's poles. These craters have been in shadow for the past two billion years, and computer simulations suggest that up to 14,000 km^2 might be in permanent shadow. The presence of usable quantities of water on the Moon is an important factor in rendering lunar habitation cost-effective, since transporting it from Earth would be prohibitively expensive. Many different signatures of lunar water have since been found. In 1994,

Clementine's bistatic radar experiment found indications of small, frozen pockets of water close to the surface (though later Arecibo radar observations suggested these might be rocks ejected from young impact craters); *Lunar Prospector*'s neutron spectrometer indicated in 1998 that high concentrations of hydrogen are present in the upper metre of the regolith near the polar regions; in 2008, new analysis found small amounts of water in the interior of volcanic lava beads brought to Earth by Apollo 15. *Chandrayaan-1*'s imaging spectrometer detected water and hydroxyl absorption lines in reflected sunlight, evidence of large quantities of water on the Moon's surface, possibly as high as 1,000 ppm. Weeks later, the *LCROSS* mission flew its 2300 kg impactor into a permanently shadowed polar crater, and detected at least 100 kg of water in the plume of ejected material. A later, more definitive, study found the amount of detected water to be 155 ± 12 kg (or "5.6 ± 2.9% by mass").

Gravity and Magnetic Fields

The gravitational field of the Moon has been measured through tracking the Doppler shift of radio signals emitted by orbiting spacecraft. The main lunar gravity features are mascons, large positive gravitational anomalies associated with some of the giant impact basins, partly caused by the dense mare basaltic lava flows that fill these basins. These anomalies greatly influence the orbit of spacecraft about the Moon. There are some puzzles: lava flows by themselves cannot explain all of the gravitational signature, and some mascons exist that are not linked to mare volcanism. The Moon has an external magnetic field of the order of one to a hundred nanoteslas, less than one-hundredth that of the Earth. It does not currently have a global dipolar magnetic field, as would be generated by a liquid metal core geodynamo, and only has crustal magnetization, probably acquired early in lunar history when a geodynamo was still operating. Alternatively, some of the remnant magnetization may be from transient magnetic fields generated during large impact events, through the expansion of an impact-generated plasma cloud in the presence of an ambient magnetic field—this is supported by the apparent location of the largest crustal magnetizations near the antipodes of the giant impact basins.

Atmosphere

The Moon has an atmosphere so tenuous as to be nearly vacuum, with a total mass of less than 10 metric tons. The surface pressure of this small mass is around 3×10^{-15} atm (0.3 nPa); it varies with the lunar day. Its sources include outgassing and sputtering, the release of atoms from the bombardment of lunar soil by solar wind ions. Elements that have been detected include sodium and potassium, produced by sputtering, which are also found in the atmospheres of Mercury and Io; helium-4 from the solar wind; and argon-40, radon-222, and polonium-210, outgassed after their creation by radioactive decay within the crust and mantle.

The absence of such neutral species (atoms or molecules) as oxygen, nitrogen, carbon, hydrogen and magnesium, which are present in the regolith, is not understood. Water vapour has been detected by *Chandrayaan-1* and found to vary with latitude, with a maximum at ~60–70 degrees; it is possibly generated from the sublimation of water ice in the regolith. These gases can either return into the regolith due to the Moon's gravity, or be lost to space: either through solar radiation pressure, or if they are ionised, by being swept away by the solar wind's magnetic field.

Seasons

The Moon's axial tilt is only 1.54°, much less than the 23.44° of the Earth. Because of this, the Moon's solar illumination varies much less with season, and topographical details play a crucial role in seasonal effects. From images taken by *Clementine* in 1994, it appears that four mountainous regions on the rim of Peary crater at the Moon's north pole remain illuminated for the entire lunar day, creating peaks of eternal light. No such regions exist at the south pole. Similarly, there are places that remain in permanent shadow at the bottoms of many polar craters, and these dark craters are extremely cold: *Lunar Reconnaissance Orbiter* measured the lowest summer temperatures in craters at the southern pole at 35 K ("238 °C), and just 26 K close to the winter solstice in north polar Hermite Crater. This is the coldest temperature in the Solar System ever measured by a spacecraft, colder even than the surface of Pluto.

RELATIONSHIP TO EARTH

Orbit

The Moon makes a complete orbit around the Earth with respect to the fixed stars about once every 27.3 days (its sidereal period). However, since the Earth is moving in its orbit about the Sun at the same time, it takes slightly longer for the Moon to show the same phase to Earth, which is about 29.5 days (its synodic period). Unlike most satellites of other planets, the Moon orbits nearer the ecliptic plane than to the planet's equatorial plane. The Moon's orbit is subtly perturbed by the Sun and Earth in many small, complex and interacting ways. For example, the plane of the Moon's orbital motion gradually rotates, which affects other aspects of lunar motion. These follow-on effects are mathematically described by Cassini's laws.

Relative Size

The Moon is exceptionally large relative to the Earth: a quarter the diameter of the planet and 1/81 its mass. It is the largest moon in the solar system relative to the size of its planet (although Charon is larger relative to the dwarf planet Pluto). However, the Earth and Moon are still considered a planet–satellite system, rather than a double-planet system, as their barycentre, the common centre of mass, is located 1,700 km (about a quarter of the Earth's radius) beneath the surface of the Earth.

Appearance from Earth

The Moon has an exceptionally low albedo, giving it a similar reflectance to coal. Despite this, it is the second brightest object in the sky after the Sun. This is partly due to the brightness enhancement of the opposition effect; at quarter phase, the Moon is only one-tenth as bright, rather than half as bright, as at full Moon. Additionally, colour constancy in the visual system recalibrates the relations between the colours of an object and its surroundings, and since the surrounding sky is comparatively dark, the sunlit Moon is perceived as a bright object. The edges of the full Moon seem as bright as the centre, with no limb darkening, due to the reflective properties of lunar soil, which reflects more light back towards the Sun than in other directions. The Moon does appear larger when close to the horizon, but this is a purely psychological effect, known as the Moon illusion, first described in the 7th century BC.

Gravitational coupling between the Moon and the bulge nearest the Moon acts as a torque on the Earth's rotation, draining angular momentum and rotational kinetic energy from the Earth's spin. In turn, angular momentum is added to the Moon's orbit, accelerating it, which lifts the Moon into a higher orbit with a longer period.

As a result, the distance between the Earth and Moon is increasing, and the Earth's spin slowing down. Measurements from lunar ranging experiments with laser reflectors left during the Apollo missions have found that the Moon's distance to the Earth increases by 38 mm per year (though this is only 0.10 ppb/ year of the radius of the Moon's orbit).

Atomic clocks also show that the Earth's day lengthens by about 15 microseconds every year, slowly increasing the rate at which UTC is adjusted by leap seconds. This tidal drag will continue until the spin of the Earth has slowed to match the orbital period of the Moon; however, long before this could happen, the Sun will have become a red giant, engulfing the Earth.

The lunar surface also experiences tides of amplitude ~10 cm over 27 days, with two components: a fixed one due to the Earth, as they are in synchronous rotation, and a varying component from the Sun. The Earth-induced component arises from libration, a result of the Moon's orbital eccentricity; if the Moon's orbit were perfectly circular, there would only be solar tides. Libration also changes the angle from which the Moon is seen, allowing about 59% of its surface to be seen from the Earth (but only half at any instant). The cumulative effects of stress built up by these tidal forces produces *moonquakes*. Moonquakes are much less common and weaker than earthquakes, although they can last for up to an hour-a significantly longer time than terrestrial earthquakes-because of the absence of water to damp out the seismic vibrations. The existence of moonquakes was an unexpected discovery from seismometers placed on the Moon by Apollo astronauts from 1969 through 1972.

Eclipses

Eclipses can only occur when the Sun, Earth, and Moon are all in a straight line. Solar eclipses occur near a new Moon, when the Moon is between the Sun and Earth. In contrast, lunar eclipses occur near a full Moon, when the Earth is between the Sun and Moon. The angular diameters of the Moon and the Sun as seen from Earth overlap in their variation, so that both total and annular solar eclipses are possible. In a total eclipse, the Moon completely covers the disc of the Sun and the solar corona becomes visible to the naked eye. Since the distance between the Moon and the Earth is very slowly increasing over time, the angular diameter of the Moon is decreasing. This means that hundreds of millions of years ago the Moon would always completely cover the Sun on solar eclipses, and no annular eclipses were possible. Likewise, about 600 million years from now (if the angular diameter of the Sun does not change), the Moon will no longer cover the Sun completely, and only annular eclipses will occur.

Because the Moon's orbit around the Earth is inclined by about 5° to the orbit of the Earth around the Sun, eclipses do not occur at every full and new Moon. For an eclipse to occur, the Moon must be near the intersection of the two orbital planes. The periodicity and recurrence of eclipses of the Sun by the Moon, and of the Moon by the Earth, is described by the saros cycle, which has a period of approximately 18 years.

As the Moon is continuously blocking our view of a half-degree-wide circular area of the sky, the related phenomenon of occultation occurs when a bright star or planet passes behind the Moon and is occulted: hidden from view. In this way, a solar eclipse is an occultation of the Sun. Because the Moon is comparatively close to the Earth, occultations of individual stars are not visible everywhere on the planet, nor at the same time. Because of the precession of the lunar orbit, each year different stars are occulted.

STUDY AND EXPLORATION

Early Studies

Understanding of the Moon's cycles was an early development of astronomy: by the 5th century BC, Babylonian astronomers had recorded the 18-year Saros cycle of lunar eclipses, and Indian astronomers had described the Moon's monthly elongation. The Chinese astronomer Shi Shen (fl. 4th century BC) gave instructions for predicting solar and lunar eclipses. Later, the physical form of the Moon and the cause of moonlight became understood. The ancient Greek philosopher Anaxagoras (d. 428 BC) reasoned that the Sun and Moon were both giant spherical rocks, and that the latter reflected the light of the former. Although the Chinese of the Han Dynasty believed the Moon to be energy equated to *qi*, their 'radiating influence' theory also recognized that the light of the Moon was merely a reflection of the Sun, and Jing Fang (78–37 BC) noted the sphericity of the Moon. In 499 AD, the Indian astronomer Aryabhata mentioned in his *Aryabhatiya* that reflected sunlight is the cause of the shining of the Moon. The astronomer and physicist Alhazen (965–1039) found that sunlight was not reflected from the Moon like a mirror, but that light was emitted from every part of the Moon's sunlit surface in all directions. Shen Kuo (1031–1095) of the Song Dynasty created an allegory equating the waxing and waning of the Moon to a round ball of reflective silver that, when doused with white powder and viewed from the side, would appear to be a crescent.

In Aristotle's (384–322 BC) description of the universe, the Moon marked the boundary between the spheres of the mutable elements (earth, water, air and fire), and the imperishable stars of aether, an influential philosophy that would dominate for centuries. However, in the 2nd century BC, Seleucus of Seleucia correctly theorized that tides were due to the attraction of the Moon, and that their height depends on the Moon's position relative to the Sun. In the same century, Aristarchus computed the size and distance of the Moon from Earth, obtaining a value of about twenty times the Earth radius for the distance. These figures were greatly improved by Ptolemy (90–168 AD): his values of a mean distance of 59 times the Earth's radius and a diameter of 0.292 Earth diameters were close to the correct values of about 60 and 0.273 respectively. Archimedes (287–212 BC) invented a planetarium calculating motions of the Moon and the known planets.

During the Middle Ages, before the invention of the telescope, the Moon was increasingly recognised as a sphere, though many believed that it was "perfectly smooth". In 1609, Galileo Galilei drew one of the first telescopic drawings of the Moon in his book *Sidereus Nuncius* and noted that it was not smooth but had mountains and craters.

Telescopic mapping of the Moon followed: later in the 17th century, the efforts of Giovanni Battista Riccioli and Francesco Maria Grimaldi led to the system of naming of lunar features in use today. The more exact 1834-6 *Mappa Selenographica* of Wilhelm Beer and Johann Heinrich Mädler, and their associated 1837 book *Der Mond*, the first trigonometrically accurate study of lunar features, included the heights of more than a thousand mountains, and introduced the study of the Moon at accuracies possible in earthly geography.

Lunar craters, first noted by Galileo, were thought to be volcanic until the 1870s proposal of Richard Proctor that they were formed by collisions. This view gained support in 1892 from the experimentation of geologist Grove Karl Gilbert, and from comparative studies from 1920 to the 1940s, leading to the development of lunar stratigraphy, which by the 1950s was becoming a new and growing branch of astrogeology.

FIRST DIRECT EXPLORATION: 1959–1976

Soviet Missions

The Cold War-inspired space race between the Soviet Union and the U.S. led to an acceleration of interest in exploration of the Moon. Once launchers had the necessary capabilities, these nations sent unmanned probes on both flyby and impact/lander missions.

Spacecraft from the Soviet Union's *Luna* program were the first to accomplish a number of goals: following three unnamed, failed missions in 1958, the first man-made object to escape Earth's gravity and pass near the Moon was *Luna 1*; the first man-made object to impact the lunar surface was *Luna 2*, and the first photographs of the normally occluded far side of the Moon were made by *Luna 3*, all in 1959.

The first spacecraft to perform a successful lunar soft landing was *Luna 9* and the first unmanned vehicle to orbit the Moon was *Luna 10*, both in 1966. Rock and soil samples were brought back to Earth by three *Luna* sample return missions (*Luna 16, 20,* and *24*), which returned 0.3 kg total. Two pioneering robotic spacecrafts of rover type landed on the Moon in 1970 and 1973 as a part of Soviet Lunokhod programme.

United States Missions

American lunar exploration began with robotic missions aimed at developing understanding of the lunar surface for an eventual manned landing: the Jet Propulsion Laboratory's *Surveyor* program landed its first spacecraft four months after *Luna 9*. NASA's manned Apollo program was developed in parallel; after a series of unmanned and manned tests of the Apollo spacecraft in Earth orbit, and spurred on by a potential Soviet lunar flight, in 1968 Apollo 8 made the first crewed mission to lunar orbit. The subsequent landing of the first humans on the Moon in 1969 is seen by many as the culmination of the space race. Neil Armstrong became the first person to walk on the Moon as the commander of the American mission Apollo 11 by first setting foot on the Moon at 02:56 UTC on 21 July 1969. The Apollo missions 11 to 17 (except Apollo 13, which aborted its planned lunar landing) returned 382 kg of lunar rock and soil in 2,196 separate samples. The American Moon landing and return was enabled by considerable technological advances in the early 1960s, in domains such as ablation chemistry, software engineering and atmospheric re-entry technology, and by highly competent management of the enormous technical undertaking.

Scientific instrument packages were installed on the lunar surface during all the Apollo missions. Long-lived instrument stations, including heat flow probes, seismometers, and magnetometers, were installed at the Apollo 12, 14, 15, 16, and 17 landing sites. Direct transmission of data to Earth concluded in late 1977 due to budgetary considerations, but as the stations' lunar laser ranging corner-cube retroreflector arrays are passive instruments, they are still being used. Ranging to the stations is routinely performed from earth-based stations with an accuracy of a few centimetres, and data from this experiment are being used to place constraints on the size of the lunar core.

Current Era: 1990–present

Post-Apollo and *Luna*, many more countries have become involved in direct exploration of the Moon. In 1990, Japan orbited the Moon with the *Hiten* spacecraft, becoming the third country to place a spacecraft into lunar orbit. The spacecraft released a smaller probe, *Hagoromo*, in lunar orbit, but the transmitter failed, preventing further scientific use of the mission. In 1994, the U.S. sent the joint Defense Department/ NASA spacecraft *Clementine* to lunar orbit. This mission obtained the first near-global topographic map of the Moon, and the first global multispectral images of the lunar surface. This was followed in 1998 by the *Lunar Prospector* mission, whose instruments indicated the presence of excess hydrogen at the lunar poles, which is likely to have been caused by the presence of water ice in the upper few meters of the regolith within permanently shadowed craters.

The European spacecraft *Smart 1*, the second ion-propelled spacecraft, was in lunar orbit from 15 November 2004 until its lunar impact on 3 September 2006, and made the first detailed survey of chemical elements on the lunar surface. China has expressed ambitious plans for exploring the Moon, and successfully orbited its first spacecraft, *Chang'e-1*, from 5 November 2007 until its controlled lunar impact on 1 March 2008. In its sixteen-month mission, it obtained a full image map of the Moon. Between 4 October 2007 and 10 June 2009, the Japan Aerospace Exploration Agency's *Kaguya (Selene)* mission, a lunar orbiter fitted with a high-definition video camera, and two small radio-transmitter satellites, obtained lunar geophysics data and took the first high-definition movies from beyond Earth orbit. India's first lunar mission, *Chandrayaan I*, orbited from 8 November 2008 until loss of contact on 27 August 2009, creating a high resolution chemical, mineralogical and photo-geological map of the lunar surface, and confirming the presence of water molecules in lunar soil. The Indian Space Research Organisation plans to launch *Chandrayaan II* in 2013, which is slated to include a Russian robotic lunar rover. The U.S. co-launched the *Lunar Reconnaissance Orbiter* (LRO) and the *LCROSS* impactor and follow-up observation orbiter on 18 June 2009; *LCROSS* completed its mission by making a planned and widely observed impact in the crater Cabeus on 9 October 2009, while *LRO* is currently in operation, obtaining precise lunar altimetry and high-resolution imagery.

Other upcoming lunar missions include Russia's *Luna-Glob*: an unmanned lander, set of seismometers, and an orbiter based on its Martian *Phobos-Grunt* mission, which is slated to launch in 2012. Privately funded lunar exploration has been promoted by the Google Lunar X Prize, announced 13 September 2007, which offers US$20 million to anyone who can land a robotic rover on the Moon and meet other specified criteria.

NASA began to plan to resume manned missions following the call by U.S. President George W. Bush on 14 January 2004 for a mission to the Moon by 2020. The Constellation program was funded and construction and testing begun on a manned spacecraft and launch vehicle, and design studies for a lunar base. However, that program has been placed in jeopardy by the proposed 2011 budget, which will cancel Constellation in favour of NASA pursuing space technology and heavy-lift rocketry research. India has also expressed its hope to send a manned mission to the Moon by 2020.

LEGAL STATUS

Although *Luna* landers scattered pennants of the Soviet Union on the Moon, and U.S. flags were symbolically planted at their landing sites by the Apollo astronauts, no nation currently claims ownership of any part of the Moon's surface. Russia and the U.S. are party to the 1967 Outer Space Treaty, which defines the Moon and all outer space as the "province of all mankind". This treaty also restricts the use of the Moon to peaceful purposes, explicitly banning military installations and weapons of mass destruction. The 1979 Moon Agreement was created to restrict the exploitation of the Moon's resources by any single nation, but it has not been signed by any of the space-faring nations. While several individuals have made claims to the Moon in whole or in part, none of these are considered credible.

In Culture

The Moon's regular phases make it a very convenient timepiece, and the periods of its waxing and waning form the basis of many of the oldest calendars. An eagle-bone tally stick, found near the village of Le Placard in France and dated to 13,000 years ago, is believed by many to mark the phases of the Moon. The ~30-day month is an approximation of the lunar cycle. The English noun *month* and its cognates in other Germanic languages stem from Proto-Germanic **mãnSth-*, which is connected to the above mentioned Proto-Germanic **mãnôn*, indicating the usage of a lunar calendar among the Germanic peoples (Germanic calendar) prior to the adoption of a solar calendar. The same Indo-European root as *moon* led to the development of Latin *measure* and *menstrual*, words which echo the Moon's importance to many ancient cultures in measuring time.

The Moon has been the subject of many works of art and literature and the inspiration for countless others. It is a motif in the visual arts, the performing arts, poetry, prose and music. A 5,000-year-old rock carving at Knowth, Ireland, may represent the Moon, which would be the earliest depiction discovered. In many prehistoric and ancient cultures, the Moon was personified as a deity or other supernatural phenomenon, and astrological views of the Moon continue to be propagated today. The contrast between the brighter highlands and darker maria create the patterns seen by different cultures as the Man in the Moon, the rabbit and the buffalo, among others. The Moon has a long association with insanity and irrationality; the words *lunacy* and *loony* are derived from the Latin name for the Moon, *Luna*. Philosophers such as Aristotle and Pliny the Elder argued that the full Moon induced insanity in susceptible individuals, believing that the brain, which is mostly water, must be affected by the Moon and its power over the tides, but the Moon's gravity is too slight to affect any single person. Even today, people insist that admissions to psychiatric hospitals, traffic accidents, homicides or suicides increase during a full Moon, although there is no scientific evidence to support such claims.

GEOLOGY OF THE MOON

The geology of the Moon (sometimes called selenology, although the latter term can refer more generally to "lunar science") is quite different from that of the Earth. The Moon lacks a significant atmosphere and any bodies of water, which eliminates erosion due to weather; it does not possess any form of plate tectonics, it has a lower gravity, and because of its small size, it cools more rapidly. The complex geomorphology of the lunar surface has been formed by a combination of processes, chief among which are impact cratering and volcanism. The Moon is a differentiated body, possessing a crust, mantle and core.

Geological studies of the Moon are based on a combination of Earth-based telescope observations, measurements from orbiting spacecraft, lunar samples, and geophysical data. A few locations were sampled directly during the Apollo missions in the late 1960s and early 1970s, which returned approximately 385 kilograms of lunar rock and soil to Earth, as well as several missions of the Soviet Luna programme. The Moon is the only extraterrestrial body for which we possess samples with a known geologic context. A handful of lunar meteorites have been recognized on Earth, though their source craters on the Moon are unknown. A substantial portion of the lunar surface has not been explored, and a number of geological questions remain unanswered.

Elemental Composition

Elements known to be present on the lunar surface include, among others, oxygen (O), silicon (Si), iron (Fe), magnesium (Mg), calcium (Ca), aluminium (Al), manganese (Mn) and titanium (Ti). Among the more abundant are oxygen, iron and silicon. The oxygen content is estimated at 45%. Carbon (C) and nitrogen (N) appear to be present only in trace quantities from deposition by solar wind.

Neutron spectrometry data from the Lunar Prospector indicate the presence of hydrogen (H) concentrated at the poles.

Formation

For a long time, the fundamental question regarding the history of the Moon was of its origin. Early hypotheses included *fission* from the Earth, *capture*, and *co-accretion*. Today, the *giant impact hypothesis* is widely accepted by the scientific community.

Fission Hypothesis

The idea that the early Earth, with an accelerated rotation, expelled a piece of its mass was proposed by George Darwin (son of the famous biologist Charles Darwin). It was commonly assumed that the Pacific Ocean represented the scar of this event. However, today it is known that the oceanic crust that makes up this ocean basin is relatively young, about 200 million years old or less, whereas the Moon is much older. This hypothesis cannot account for the angular momentum of the Earth-Moon system.

Lunar Capture

This hypothesis states that the Moon was captured, completely formed, by the gravitational field of the Earth. This is unlikely, since a close encounter with the Earth would have produced either a collision or an alteration of the trajectory of the body in question, so if it had indeed happened, the Moon probably would never return to meet again with the Earth. For this hypothesis to function, there would have to be a large atmosphere extended around the primitive Earth, which would be able to slow the movement of the Moon before it could escape. This hypothesis is considered to explain the irregular satellite orbits of Jupiter and Saturn. In addition, this hypothesis has difficulty explaining the similar oxygen isotope ratio of the two worlds.

Co-accretion Hypothesis

This hypothesis states that the Earth and the Moon formed together as a double system from the primordial accretion disk of the Solar System. The problem with this hypothesis is that it does not explain the angular momentum of the Earth-Moon system, nor why the Moon has a relative small iron core compared to the Earth (25% of its radius compared to 50% for the Earth).

Giant Impact Theory

At present the best explanation for the origin of the Moon involves a collision of two protoplanetary bodies during the early accretional period of Solar System evolution. This "giant impact theory", which became popular in 1984 (although Reginald Aldworth Daly, a Canadian professor at Harvard college, originated it in the 1940s), satisfies the orbital conditions of the Earth and Moon and can account for the relatively small metallic core of the Moon. Collisions between planetesimals are now recognized to lead to the growth of planetary bodies early in the evolution of the Solar System, and in this framework it is inevitable that large impacts will sometimes occur when the planets are nearly formed.

The theory requires a collision between a body about 90% the present size of the Earth, and another the diameter of Mars (half of the terrestrial radius and a tenth of its mass). The colliding body has sometimes been referred to as Theia, the mother of Selene, the Moon goddess in Greek mythology. This size ratio is needed in order for the resulting system to possess sufficient angular momentum to match the current orbital configuration. Such an impact would have put enough material into orbit about the Earth to have eventually accumulated to form the Moon.

Computer simulations of this event appear to show that the collision must occur with a somewhat glancing blow. This will cause a small portion of the colliding body to form a long arm of material that will then shear off.

The asymmetrical shape of the Earth following the collision then causes this material to settle into an orbit around the main mass. The energy involved in this collision is impressive: trillions of tons of material would have been vaporized and melted. In parts of the Earth the temperature would have risen to 10,000°C (18,000°F).

This formation theory helps explain why the Moon possesses only a small iron core (roughly 25% of its radius, in comparison to about 50% for the Earth). Most of the iron core from the impacting body is predicted to have accreted to the core of the Earth. The lack of volatiles in the lunar samples is also explained in part by the energy of the collision. The energy liberated during the reaccreation of material in orbit about the Earth would have been sufficient to melt a large portion of Moon, leading to the generation of a magma ocean.

The newly formed moon orbited at about one-tenth the distance that it does today, and became tidally-locked with the Earth, where one side continually faces toward the Earth. The geology of the Moon has since been independent of the Earth. While this theory explains many aspects of the Earth-Moon system, there are still a few unresolved problems facing this theory, such as the Moon's volatile elements not being as depleted as expected from such an energetic impact.

Geologic History

The geological history of the Moon has been defined into six major epochs, called the lunar geologic timescale. Starting about 4.5 billion years ago, the newly formed Moon was in a molten state and was orbiting much closer to the Earth. The resulting tidal forces deformed the molten body into an ellipsoid, with the major axis pointed towards Earth.

The first important event in the geologic evolution of the Moon was the crystallization of the near global magma ocean. It is not known with certainty what its depth was, but several studies imply a depth of about 500 km or greater. The first minerals to form in this ocean were the iron and magnesium silicates olivine and pyroxene. Because these minerals were denser than the molten material around them, they sank. After crystallization was about 75% complete, less dense anorthositic plagioclase feldspar crystallized and floated, forming an anorthositic crust about 50 km in thickness. The majority of the magma ocean crystallized quickly (within about 100 million years or less), though the final remaining KREEP-rich magmas, which are highly enriched in incompatible and heat producing elements, could have remained partially molten for several hundred million (or perhaps 1 billion) years. It appears that the final KREEP-rich magmas of the magma ocean eventually became concentrated within the region of Oceanus Procellarum and the Imbrium basin, a unique geologic province that is now known as the Procellarum KREEP Terrane.

Quickly after the lunar crust formed, or even as it was forming, different types of magmas that would give rise to the Mg-suite norites and troctolites began to form, although the exact depths at which this occurred are not known precisely. Recent theories suggest that Mg-suite plutonism was largely confined to the region of the Procellarum KREEP Terrane, and that these magmas are genetically related to KREEP in some manner, though their origin is still highly debated in the scientific community. The oldest of the Mg-suite rocks have crystallization ages of about 3.85 Ga. However, the last large impact that could have excavated deep into the crust (the Imbrium basin) also occurred at 3.85 Ga before present. Thus, it seems probable that Mg-suite plutonic activity continued for a much longer time, and that younger plutonic rocks exist deep below the surface. Analysis of the lunar samples seem to imply that a significant percentage of the lunar impact basins formed within a very short period of time between about 4 and 3.85 Ga ago. This hypothesis is referred to as the lunar cataclysm or late heavy bombardment. However, it is now recognized that ejecta from the Imbrium impact basin (one of the youngest large impact basins on the Moon) should be found at all of the Apollo landing sites. It is thus possible that ages for some impact basins (in particular Mare Nectaris) could have been mistakenly assigned the same age as Imbrium.

The lunar maria represent ancient flood basaltic eruptions. In comparison to terrestrial lavas, these contain higher iron abundances, have low viscosities, and some contain highly elevated abundances of the titanium-rich mineral ilmenite. The majority of basaltic eruptions occurred between about 3 and 3.5 Ga ago, though some mare samples have ages as old as 4.2 Ga, and the youngest (based on the method of crater counting) are believed to have erupted only 1 billion years ago. Along with mare volcanism came pyroclastic eruptions, which launched molten basaltic materials hundreds of kilometres away from the volcano. A large portion of the mare formed, or flowed into, the low elevations associated with the nearside impact basins. However, it must be noted that Oceanus Procellarum does not correspond to any known impact structure, and the lowest elevations of the Moon within the farside South Pole-Aitken basin are only modestly covered by mare.

Impacts by meteorites and comets are the only abrupt geologic force acting on the Moon today, though the variation of Earth tides on the scale of the Lunar anomalistic month causes small variations in stresses. Some of the most important craters used in lunar stratigraphy formed in this recent epoch. For example, the crater Copernicus, which has a depth of 3.76 km and a radius of 93 km, is believed to have formed about 900 million years ago (though this is debatable). The *Apollo 17* mission landed in an area in which the material coming from the crater Tycho might have been sampled. The study of these rocks seem to indicate that this crater could have formed 100 million years ago, though this is debatable as well. The surface has also experienced space weathering due to high energy particles, solar wind implantation, and micrometeorite impacts.

This process causes the ray systems associated with young craters to darken until it matches the albedo of the surrounding surface. However, if the composition of the ray is different from the underlying crustal materials (as might occur when a "highland" ray is emplaced on the mare), the ray could be visible for much longer times.

After resumption of Lunar exploration in the 1990s, it was discovered there are scarps across the globe which are caused by the contraction due to cooling of the Moon.

LUNAR LANDSCAPE

The lunar landscape is characterized by impact craters, their ejecta, a few volcanoes, hills, lava flows and depressions filled by magma.

The Highlands

The most distinctive aspect of the Moon is the contrast between its light and dark zones. Lighter surfaces are the lunar highlands, which receive the name of *terrae* (singular *terra*, from the Latin for Earth), and the darker plains are called *maria* (singular *mare*, from the Latin for sea), after Johannes Kepler who introduced the name in the 17th century. The highlands are anorthositic in composition, whereas the maria are basaltic. The maria often coincide with the "lowlands," but it is important to note that the lowlands (such as within the South Pole-Aitken basin) are not always covered by maria. The highlands are older than the visible maria, and hence are more heavily cratered.

The Maria

The major products of volcanic processes on the Moon are evident to the Earth-bound observer in the form of the lunar maria. These are large flows of basaltic lava that correspond to low-albedo surfaces covering nearly a third of the near side. Only a few percent of the farside has been affected by mare volcanism. Even before the Apollo missions confirmed it, most scientists believed that the maria were lava-filled plains, since they possessed lava flow patterns and collapses attributed to lava tubes.

The ages of the mare basalts have been determined both by direct radiometric dating and by the technique of crater counting. The oldest radiometric ages are about 4.2 Ga, whereas the youngest ages determined from crater counting are about 1 Ga (1 Ga = 1 billion years). Volumetrically, most of the mare formed between about 3 and 3.5 Ga before present. The youngest lavas erupted within Oceanus Procellarum, whereas some of the oldest appear to be located on the farside. The maria are clearly younger than the surrounding highlands given their lower density of impact craters.

A large portion of maria erupted within, or flowed into, the low-lying impact basins on the lunar nearside. Nevertheless, it is unlikely that a causal relationship exists between the impact event and mare volcanism because the impact basins are much older (by about 500 million years) than the mare fill. Furthermore, Oceanus Procellarum, which is the largest expanse of mare volcanism on the Moon, does not correspond to any known impact basin. It is commonly suggested that the reason the mare only erupted on the nearside is that the nearside crust is thinner than the farside. While crustal thickness variations might act to modulate the amount of magma that ultimately reaches the surface, this hypothesis does not explain why the farside South Pole-Aitken basin, whose crust is thinner than Oceanus Procellarum, was only modestly filled by volcanic products. Finally, it should be noted that the Earth's gravity played no preferential role in causing mare volcanism to occur on the near side, as the Earth's gravitational attraction is exactly balanced by the centrifugal acceleration resulting from the Moon's rotation.

Another type of deposit associated with the maria, although it also covers the highland areas, are the "dark mantle" deposits. These deposits cannot be seen with the naked eye, but they can be seen in images taken from telescopes or orbiting spacecraft. Before the Apollo missions, scientists believed that they were deposits produced by pyroclastic eruptions. Some deposits appear to be associated with dark elongated ash cones, reinforcing the idea of pyroclasts. The existence of pyroclastic eruptions was later confirmed by the discovery of glass spherules similar to those found in pyroclastic eruptions here on Earth.

Many of the lunar basalts contain small holes called vesicles, which were formed by gas bubbles exsolving from the magma at the vacuum conditions encountered at the surface. It is not known with certainty which gases escaped these rocks, but carbon monoxide is one candidate.

The samples of pyroclastic glasses are of green, yellow, and red tints. The difference in colour indicates the concentration of titanium that the rock possesses, with the green particles having the lowest concentrations (about 1%), and red particles having the highest concentrations (up to 14%, much more than the basalts with the highest concentrations).

Rilles

Rilles on the Moon sometimes resulted from the formation of localized lava channels. These generally fall into three categories, consisting of sinuous, arcuate, or linear shapes. By following these meandering rilles back to their source, they often lead to an old volcanic vent. One of the most notable sinuous rilles is the Vallis Schröteri feature, located in the Aristarchus plateau along the eastern edge of Oceanus Procellarum. An example of a sinuous rille exists at the Apollo 15 landing site, Rima Hadley, located on the rim of the Imbrium Basin. Based on observations from the mission, it is generally believed that this rille was formed by volcanic processes, a topic long debated before the mission took place.

Domes

A variety of shield volcanoes can be found in selected locations on the lunar surface, such as on Mons Rümker. These are believed to be formed by relatively viscous, possibly silica-rich lava, erupting from localized vents. The resulting lunar domes are wide, rounded, circular features with a gentle slope rising in elevation a few hundred meters to the mid-point. They are typically 8–12 km in diameter, but can be up to 20 km across. Some of the domes contain a small pit at their peak.

Wrinkle Ridges

Wrinkle ridges are features created by compressive tectonic forces within the maria. These features represent buckling of the surface and form long ridges across parts of the maria. Some of these ridges may outline buried craters or other features beneath the maria. A prime example of such an outlined feature is the crater Letronne.

Grabens

Grabens are tectonic features that form under extension stresses. Structurally, they are composed of two normal faults, with a down-dropped block between them. Most grabens are found within the lunar maria near the edges of large impact basins.

Impact Craters

It may be surprising to learn that the origin of the Moon's craters as impact features became widely accepted only in the 1940s. This realization allowed the impact history of the Moon to be gradually worked out by means of the geologic principle of superposition. That is, if a crater (or its ejecta) overlaid another, it must be the younger. The amount of erosion experienced by a crater was another clue to its age, though this is more subjective. Adopting this approach in the late 1950s, Gene Shoemaker took the systematic study of the Moon away from the astronomers and placed it firmly in the hands of the lunar geologists.

Impact cratering is the most notable geological process on the Moon. The craters are formed when a solid body, such as an asteroid or comet, collides with the surface at a high velocity (mean impact velocities for the Moon are about 17 km per second). The kinetic energy of the impact creates a compression shock wave that radiates away from the point of entry. This is succeeded by a rarefaction wave, which is responsible for propelling ı st of the ejecta out of the crater. Finally there is a hydrodynamic rebound of the floor that can create a central peak.

These craters appear in a continuum of diameters across the surface of the Moon, ranging in size from tiny pits to the immense South Pole-Aitken Basin with a diameter of nearly 2,500 km and a depth of 13 km. In a very general sense, the lunar history of impact cratering follows a trend of decreasing crater size with time. In particular, the largest impact basins were formed during the early periods, and these were successively overlaid by smaller craters. The size frequency distribution (SFD) of crater diameters on a given surface (that is, the number of craters as a function of diameter) approximately follows a power law with increasing number of craters with decreasing crater size. The vertical position of this curve can be used to estimate the age of the surface. The most recent impacts are distinguished by well-defined features, including a sharp-edged rim. Small craters tend to form a bowl shape, while larger impacts can have a central peak with flat floors. Larger craters generally display slumping features along the inner walls that can form terraces and ledges. The largest impact basins, the multiring basins, can even have secondary concentric rings of raised material.

The impact process excavates high albedo materials that initially gives the crater, ejecta, and ray system a bright appearance. The process of space weathering gradually decreases the albedo of this material such that the rays fade with time. Gradually the crater and its ejecta undergo impact erosion from micrometeorites and smaller impacts. This erosional process softens and rounds the features of the crater. The crater can also be covered in ejecta from other impacts, which can submerge features and even bury the central peak. The ejecta from large impacts can include larges blocks of material that reimpact the surface to form secondary impact craters. These craters are sometimes formed in clearly discernible radial patterns, and generally have shallower depths than primary craters of the same size. In some cases an entire line of these blocks can impact to form a valley. These are distinguished from *catena*, or crater chains, which are linear strings of craters that are formed when the impact body breaks up prior to impact.

Generally speaking, a lunar crater is roughly circular in form. Laboratory experiments at NASA's Ames Research Centre have demonstrated that even very low-angle impacts tend to produce circular craters, and that elliptical craters start forming at impact angles below five degrees. However, a low angle impact can produce a central peak that is offset from the mid-point of the crater. Additionally, the ejecta from oblique impacts show distinctive patterns at different impact angles: asymmetry starting around 60° and a wedge-shaped "zone of avoidance" free of ejecta in the direction the projectile came from starting around 45°.

Dark-halo craters are formed when an impact excavates lower albedo material from beneath the surface, then deposits this darker ejecta around the main crater. This can occur when an area of darker basaltic material, such as that found on the maria, is later covered by lighter ejecta derived from more distant impacts in the highlands. This covering conceals the darker material below, which is later excavated by subsequent craters.

The largest impacts produced melt sheets of molten rock that covered portions of the surface which could be as thick as a kilometre. Examples of such impact melt can be seen in the northeastern part of the Mare Orientale impact basin.

Regolith

The surface of the Moon has been subject to billions of years of collisions with both small and large asteroidal and cometary materials. Over time, these impact processes have pulverized and "gardened" the surface materials, forming a fine grained layer termed "regolith".

The thickness of the regolith varies between 2 meters beneath the younger maria, to up to 20 meters beneath the oldest surfaces of the lunar highlands. The regolith is predominantly composed of materials found in the region, but also contains traces of materials ejected by distant impact craters. The term "mega-regolith" is often used to describe the heavily fractured bedrock directly beneath the near-surface regolith layer.

The regolith contains rocks, fragments of minerals from the original bedrock, and glassy particles formed during the impacts. In most of the lunar regolith, half of the particles are made of mineral fragments fused by the glassy particles; these objects are called agglutinates. The chemical composition of the regolith varies according to its location; the regolith in the highlands is rich in aluminium and silica, just as the rocks in those regions. The regolith in the maria is rich in iron and magnesium and is silica-poor, as the basaltic rocks from which it is formed.

The lunar regolith is very important because it also stores information about the history of the Sun. The atoms that compose the solar wind – mostly helium, neon, carbon and nitrogen – hit the lunar surface and insert themselves into the mineral grains. Upon analyzing the composition of the regolith, particularly its isotopic composition, it is possible to determine if the activity of the Sun has changed with time.

The gases of the solar wind could be useful for future lunar bases, since oxygen, hydrogen (water), carbon and nitrogen are not only essential to sustain life, but are also potentially very useful in the production of fuel. The composition of the lunar regolith can also be used to infer its source origin.

Lunar Lava Tubes

Lunar lava tubes form a potentially important location for constructing a future lunar base, which may be used for local exploration and development, or as a human outpost to serve exploration beyond the Moon.

Any intact lava tube on the moon could serve as a shelter from the severe environment of the lunar surface, with its frequent meteorite impacts, high-energy ultraviolet radiation and energetic particles, and extreme diurnal temperature variations.. Following the launch of the Lunar Reconnaissance Orbiter, many lunar lava tubes have been imaged. These lunar pits are found in several locations across the moon, including Marius Hills, Mare Ingenii and Mare Tranquillitatis.

THE LUNAR MAGMA OCEAN

The first rocks brought back by Apollo 11 were basalts. Although the mission landed on Mare Tranquillitatis, a few millimetric fragments of rocks coming from the highlands were picked up. These are composed mainly of plagioclase feldspar; some fragments were composed exclusively of anorthositic plagioclase. The identification of these mineral fragments led to the bold hypothesis that a large portion of the Moon was once molten, and that the crust formed by fractional crystallization of this magma ocean.

A natural outcome of the giant impact event is that the materials that reaccreted to form the Moon must have been hot. Current models predict that a large portion of the Moon would have been molten shortly after the Moon formed, with estimates for the depth of this magma ocean ranging from about 500 km to full moon melting. Crystallization of this magma ocean would have given rise to a differentiated body with a compositionally distinct crust and mantle and accounts for the major suites of lunar rocks.

As crystallization of the lunar magma ocean proceeded, minerals such as olivine and pyroxene would have precipitated and sank to form the lunar mantle. After crystallization was about three-quarters complete, anorthositic plagioclase would have begun to crystallize, and because of its low density, float, forming an anorthositic crust. Importantly, elements that are incompatible (i.e., those that partition preferentially into the liquid phase) would have been progressively concentrated into the magma as crystallization progressed, forming a KREEP-rich magma that initially should have been sandwiched between the crust and mantle.

Evidence for this scenario comes from the highly anorthositic composition of the lunar highland crust, as well as the existence of KREEP-rich materials.

LUNAR ROCKS

Surface Materials

The Apollo program brought back 381.7 kg (841.5 lb) of lunar surface material, most of which is stored at the Lunar Receiving Laboratory in Houston, Texas. These rocks have proved to be invaluable in deciphering the geologic evolution of the Moon.

Lunar rocks are in large part made of the same common rock forming minerals as found on Earth, such as olivine, pyroxene, and plagioclase feldspar (anorthosite). Plagioclase feldspar is mostly found in the lunar crust, while pyroxene and olivine are typically seen in the lunar mantle. The mineral ilmenite is highly abundant in some mare basalts, and a new mineral named armalcolite (named for *Arm*strong, *Al*drin, and *Col*lins, the three members of the Apollo 11 crew) was first discovered in the lunar samples.

The maria are composed predominantly of basalt, whereas the highland regions are iron-poor and composed primarily of anorthosite, a rock composed primarily of calcium-rich plagioclase feldspar. Another significant component of the crust are the igneous *Mg-suite* rocks, such as the troctolites, norites, and KREEP-basalts. These rocks are believed to be genetically related to the petrogenesis of KREEP. Composite rocks on the lunar surface often appear in the form of breccias. Of these, the subcategories are called fragmental, granulitic, and impact-melt breccias, depending on how they were formed. The mafic impact melt breccias, which are typified by the *low-K Fra Mauro* composition, have a higher proportion of iron and magnesium than typical upper crust anorthositic rocks, as well as higher abundances of KREEP.

Composition of the Maria

The main characteristics of the basaltic rocks with respect to the rocks of the lunar highlands is that the basalts contain higher abundances of olivine and pyroxene, and less plagioclase. They are more rich in iron than terrestrial basalts, and also have lower viscosities.

Some of them have high abundances of a ferro-titanic oxide called ilmenite. Since the first sampling of rocks contained a high content of ilmenite and other related minerals, they received the name of "high titanium" basalts.

The Apollo 12 mission returned to Earth with basalts of lower titanium concentrations, and these were dubbed "low titanium" basalts. Subsequent missions, including the Soviet unmanned probes, returned with basalts with even lower concentrations, now called "very low titanium" basalts. The Clementine space probe returned data showing that the mare basalts possess a continuum in titanium concentrations, with the highest concentration rocks being the least abundant.

Internal Structure of the Moon

The current model of the interior of the Moon was derived using seismometers left behind during the manned Apollo program missions, as well as investigations of the Moon's gravity field and rotation. The mass of the Moon is sufficient to eliminate any voids within the interior, so it is believed to be composed of solid rock throughout. Its low bulk density (~3346 kg m^{-3}) indicates a low metal abundance.

Mass and moment of inertia constraints indicate that the Moon likely has an iron core that is less than about 450 km in radius. Studies of the Moon's physical librations (small perturbations to its rotation) furthermore indicate that the core is still molten. Most planetary bodies and moons have iron cores that are about half the size of the body. The Moon is thus anomalous in possessing a core whose size is only about one quarter of its radius.

The crust of the Moon is on average about 50 km thick (though this is uncertain by about ±15 km). It is widely believed that the far-side crust is on average thicker than the near side by about 15 km. Seismology has constrained the thickness of the crust only near the Apollo 12 and 14 landing sites. While the initial Apollo-era analyses suggested a crustal thickness of about 60 km at this site, recent reanalyses of this data set suggest a thinner value, somewhere between about 30 and 45 km.

Compared to that of Earth, the Moon has only a very weak external magnetic field. Other major differences are that the Moon does not currently have a dipolar magnetic field (as would be generated by a geodynamo in its core), and the magnetizations that are present are almost entirely crustal in origin. One hypothesis holds that the crustal magnetizations were acquired early in lunar history when a geodynamo was still operating.

The small size of the lunar core, however, is a potential obstacle to this theory. Alternatively, it is possible that on airless bodies such as the Moon, transient magnetic fields could be generated during impact processes. In support of this, it has been noted that the largest crustal magnetizations appear to be located near the antipodes of the largest impact basins. While the Moon does not possess a dipolar magnetic field like the Earth does, some of the returned rocks possess strong magnetizations. Furthermore measurements from orbit show that some portions of the lunar surface are associated with strong magnetic fields.

TOPOGRAPHY OF THE MOON

The topography of the Moon has been measured by the methods of laser altimetry and stereo image analysis, most recently from data obtained during the Clementine mission. The most visible topographic feature is the giant far side South Pole-Aitken basin, which possesses the lowest elevations of the Moon. The highest elevations are found just to the north-east of this basin, and it has been suggested that this area might represent thick ejecta deposits that were emplaced during an oblique South Pole-Aitken basin impact event. Other large impact basins, such as the maria Imbrium, Serenitatis, Crisium, Smythii, and Orientale, also possess regionally low elevations and elevated rims.

Another distinguishing feature of the Moon's shape is that the elevations are on average about 1.9 km higher on the far side than the near side. If it is assumed that the crust is in isostatic equilibrium, and that the density of the crust is everywhere the same, then the higher elevations would be associated with a thicker crust. Using gravity, topography and seismic data, the crust is thought to be on average about 50 ± 15 km thick, with the far-side crust being on average thicker than the near side by about 15 km.

GRAVITATION OF THE MOON

The gravitational field of the Moon has been determined by the tracking of radio signals emitted by orbiting spacecraft. The principle used depends on the Doppler effect, whereby the line-of-sight spacecraft acceleration can be measured by small shifts in frequency of the radio signal, and the measurement of the distance from the spacecraft to a station on Earth. Since the gravitational field of the Moon affects the orbit of a spacecraft, it is possible to use these tracking data to invert for gravity anomalies.

However, because of the Moon's synchronous rotation it is not possible to track spacecraft much over the limbs of the Moon, and the far-side gravity field is thus only poorly characterized. The gravitational acceleration on the surface of the moon is 1.63 m/s^2, about 16.7% that on Earth's surface. Because weight is directly dependent upon gravitational acceleration, things on the moon will weigh only 16.7% of what they weigh on the earth.

The major characteristic of the Moon's gravitational field is the presence of mascons, which are large positive gravity anomalies associated with some of the giant impact basins. These anomalies greatly influence the orbit of spacecraft about the Moon, and an accurate gravitational model is necessary in the planning of both manned and unmanned missions. They were initially discovered by the analysis of Lunar Orbiter tracking data, since navigation tests prior to the Apollo program experienced positioning errors much larger than mission specifications.

The origin of mascons are in part due to the presence of dense mare basaltic lava flows that fill some of the impact basins. However, lava flows by themselves can not explain the entirety of the gravitational variations, and uplift of the crust-mantle interface is required as well. Based on Lunar Prospector gravitational models, it has been suggested that some mascons exist that do not show evidence for mare basaltic volcanism. The huge expanse of mare basaltic volcanism associated with Oceanus Procellarum does not possess a positive gravity anomaly.

11

The Sun

The Sun is the star at the centre of the Solar System. It has a diameter of about 1,392,000 km, about 109 times that of Earth, and its mass (about 2×10^{30} kilograms, 330,000 times that of Earth) accounts for about 99.86% of the total mass of the Solar System. About three quarters of the Sun's mass consists of hydrogen, while the rest is mostly helium. Less than 2% consists of heavier elements, including oxygen, carbon, neon, iron, and others.

The Sun's colour is white, although from the surface of the Earth it may appear yellow because of atmospheric scattering of blue light. Its stellar classification, based on spectral class, is G2V, and is informally designated a *yellow star*, because its visible radiation is most intense in the yellow-green portion of the spectrum. In this spectral class label, *G2* indicates its surface temperature of approximately 5778 K (5505 °C), and *V* indicates that the Sun, like most stars, is a main sequence star, and thus generates its energy by nuclear fusion of hydrogen nuclei into helium. In its core, the Sun fuses 430–600 million tons of hydrogen each second. Once regarded by astronomers as a small and relatively insignificant star, the Sun is now thought to be brighter than about 85% of the stars in the Milky Way galaxy, most of which are red dwarfs. The absolute magnitude of the Sun is +4.83; however, as the star closest to Earth, the Sun is the brightest object in the sky with an apparent magnitude of d26.74. The Sun's hot corona continuously expands in space creating the solar wind, a stream of charged particles that extends to the heliopause at roughly

100 astronomical units. The bubble in the interstellar medium formed by the solar wind, the heliosphere, is the largest continuous structure in the Solar System.

The Sun is currently traveling through the Local Interstellar Cloud in the Local Bubble zone, within the inner rim of the Orion Arm of the Milky Way galaxy. Of the 50 nearest stellar systems within 17 light-years from Earth (the closest being a red dwarf named Proxima Centauri at approximately 4.2 light years away), the Sun ranks 4th in mass. The Sun orbits the centre of the Milky Way at a distance of approximately 24,000–26,000 light years from the galactic centre, completing one clockwise orbit, as viewed from the galactic north pole, in about 225–250 million years. Since our galaxy is moving with respect to the cosmic microwave background radiation (CMB) in the direction of constellation Hydra with a speed of 550 km/s, the sun's resultant velocity with respect to the CMB is about 370 km/s in the direction of Crater or Leo.

The mean distance of the Sun from the Earth is approximately 149.6 million kilometres (1 AU), though the distance varies as the Earth moves from perihelion in January to aphelion in July. At this average distance, light travels from the Sun to Earth in about 8 minutes and 19 seconds. The energy of this sunlight supports almost all life on Earth by photosynthesis, and drives Earth's climate and weather. The enormous effect of the Sun on the Earth has been recognized since prehistoric times, and the Sun has been regarded by some cultures as a deity. An accurate scientific understanding of the Sun developed slowly, and as recently as the 19th century prominent scientists had little knowledge of the Sun's physical composition and source of energy. This understanding is still developing; there are a number of present-day anomalies in the Sun's behavior that remain unexplained.

CHARACTERISTICS

The Sun is a G-type main sequence star comprising about 99.8632% of the total mass of the Solar System. It is a near-perfect sphere, with an oblateness estimated at about 9 millionths, which means that its polar diameter differs from its equatorial diameter by only 10 km. As the Sun consists of a plasma and is not solid, it rotates faster at its equator than at its poles. This behavior is known as differential rotation, and is caused by convection in the Sun and the movement of mass, due to steep temperature gradients from the core outwards. This mass carries a portion of the Sun's counter-clockwise angular momentum, as viewed from the ecliptic north pole, thus redistributing the angular velocity. The period of this *actual rotation* is approximately 25.6 days at the equator and 33.5 days at the poles. However, due to our constantly changing vantage point from the Earth as it orbits the Sun, the *apparent rotation* of the star at its equator is about 28 days. The centrifugal effect of this slow rotation is 18 million times weaker than the surface gravity at the Sun's equator. The tidal effect of the planets is even weaker, and does not significantly affect the shape of the Sun.

The Sun is a Population I, or heavy element-rich, star. The formation of the Sun may have been triggered by shockwaves from one or more nearby supernovae. This is suggested by a high abundance of heavy elements in the Solar System, such as gold and uranium, relative to the abundances of these elements in so-called Population II (heavy element-poor) stars. These elements could most plausibly have been produced by endergonic nuclear reactions during a supernova, or by transmutation through neutron absorption inside a massive second-generation star. The Sun does not have a definite boundary as rocky planets do, and in its outer parts the density of its gases drops exponentially with increasing distance from its centre. Nevertheless, it has a well-defined interior structure, described below. The Sun's radius is measured from its centre to the edge of the photosphere. This is simply the layer above which the gases are too cool or too thin to radiate a significant amount of light, and is therefore the surface most readily visible to the naked eye.

The solar interior is not directly observable, and the Sun itself is opaque to electromagnetic radiation. However, just as seismology uses waves generated by earthquakes to reveal the interior structure of the Earth, the discipline of helioseismology makes use of pressure waves (infrasound) traversing the Sun's interior to measure and visualize the star's inner structure. Computer modeling of the Sun is also used as a theoretical tool to investigate its deeper layers.

Core

The core of the Sun is considered to extend from the centre to about 0.2 to 0.25 of the solar radius. It has a density of up to 150 g/cm (about 150 times the density of water) and a temperature of close to 13.6 million kelvin (K). By contrast, the Sun's surface temperature is approximately 5,800 K. Recent analysis of SOHO mission data favors a faster rotation rate in the core than in the rest of the radiative zone. Through most of the Sun's life, energy is produced by nuclear fusion through a series of steps called the p–p (proton–proton) chain; this process converts hydrogen into helium. Less than 2% of the helium generated in the Sun comes from the CNO cycle.

The core is the only region in the Sun that produces an appreciable amount of thermal energy through fusion; inside 24% of the Sun's radius, 99% of the power has been generated, and by 30% of the radius, fusion has stopped nearly entirely. The rest of the star is heated by energy that is transferred outward from the core and the layers just outside. The energy produced by fusion in the core must then travel through many successive layers to the solar photosphere before it escapes into space as sunlight or kinetic energy of particles.

The proton–proton chain occurs around 9.2×10^{37} times each second in the core of the Sun. Since this reaction uses four free protons (hydrogen nuclei), it converts about 3.7×10^{38} protons to alpha particles (helium nuclei) every second (out of a total of $\sim 8.9 \times 10^{56}$ free protons in the Sun), or about 6.2×10^{11} kg per second. Since fusing hydrogen into helium releases around 0.7% of the fused mass as energy, the Sun releases energy at the mass-energy conversion rate of 4.26 million metric tons per second, 384.6 yottawatts (3.846×10^{26} W), or 9.192×10^{10} megatons of TNT per second. This mass is not destroyed to create the energy, rather, the mass is carried away *in* the radiated energy, as described by the concept of mass-energy equivalence.

The power production by fusion in the core varies with distance from the solar centre. At the centre of the Sun, theoretical models estimate it to be approximately 276.5 watts/m^3, a power production density that more nearly approximates reptile metabolism than a thermonuclear bomb. Peak power production in the Sun has been compared to the volumetric heats generated in an active compost heap. The tremendous power output of the Sun is not due to its high power per volume, but instead due to its large size.

The fusion rate in the core is in a self-correcting equilibrium: a slightly higher rate of fusion would cause the core to heat up more and expand slightly against the weight of the outer layers, reducing the fusion rate and correcting the perturbation; and a slightly lower rate would cause the core to cool and shrink slightly, increasing the fusion rate and again reverting it to its present level. The gamma rays (high-energy photons) released in fusion reactions are absorbed in only a few millimeters of solar plasma and then re-emitted again in random direction and at slightly lower energy. Therefore it takes a long time for radiation to reach the Sun's surface. Estimates of the photon travel time range between 10,000 and 170,000 years.

After a final trip through the convective outer layer to the transparent surface of the photosphere, the photons escape as visible light. Each gamma ray in the Sun's core is converted into several million photons of visible light before escaping into space. Neutrinos are also released by the fusion reactions in the core, but unlike photons they rarely interact with matter, so almost all are able to escape the Sun immediately. For many years measurements of the number of neutrinos produced in the Sun were lower than theories predicted by a factor of 3. This discrepancy was resolved in 2001 through the discovery of the effects of neutrino oscillation: the Sun emits the number of neutrinos predicted by the theory, but neutrino detectors were missing $^{2}D_{3}$ of them because the neutrinos had changed flavor by the time they were detected.

Radiative Zone

From about 0.25 to about 0.7 solar radii, solar material is hot and dense enough that thermal radiation is sufficient to transfer the intense heat of the core outward. This zone is free of thermal convection; while the material gets cooler from 7 to about 2 million kelvin with increasing altitude, this temperature gradient is less than the value of the adiabatic lapse rate and hence cannot drive convection. Energy is transferred by radiation—ions of hydrogen and helium emit photons, which travel only a brief distance before being reabsorbed by other ions. The density drops a hundredfold (from 20 g/cm^3 to only 0.2 g/cm^3) from the bottom to the top of the radiative zone.

The radiative zone and the convection form a transition layer, the tachocline. This is a region where the sharp regime change between the uniform rotation of the radiative zone and the differential rotation of the convection zone results in a large shear—a condition where successive horizontal layers slide past one another. The fluid motions found in the convection zone above, slowly disappear from the top of this layer to its bottom, matching the calm characteristics of the radiative zone on the bottom. Presently, it is hypothesized, that a magnetic dynamo within this layer generates the Sun's magnetic field.

Convective Zone

In the Sun's outer layer, from its surface down to approximately 200,000 km (or 70% of the solar radius), the solar plasma is not dense enough or hot enough to transfer the thermal energy of the interior outward through radiation; in other words it is opaque enough. As a result, thermal convection occurs as thermal columns carry hot material to the surface (photosphere) of the Sun. Once the material cools off at the surface, it plunges downward to the base of the convection zone, to receive more heat from the top of the radiative zone. At the visible surface of the Sun, the temperature has dropped to 5,700 K and the density to only 0.2 g/m^3 (about 1/6,000th the density of air at sea level). The thermal columns in the convection zone form an imprint on the surface of the Sun as the solar granulation and supergranulation. The turbulent convection of this outer part of the solar interior causes a "small-scale" dynamo that produces magnetic north and south poles all over the surface of the Sun. The Sun's thermal columns are Bénard cells and therefore tend to be hexagonal prisms.

Photosphere

The visible surface of the Sun, the photosphere, is the layer below which the Sun becomes opaque to visible light. Above the photosphere visible sunlight is free to propagate into space, and its energy escapes the Sun entirely. The change in opacity is due to the decreasing amount of H^d ions, which absorb visible light easily. Conversely, the visible light we see is produced as electrons react with hydrogen atoms to produce H^d ions. The photosphere is tens to hundreds of kilometres thick, being slightly less opaque than air on Earth. Because the upper part of the photosphere is cooler than the lower part, an image of the Sun appears brighter in the centre than on the edge or *limb* of the solar disk, in a phenomenon known as limb darkening. Sunlight has approximately a black-body spectrum that indicates its temperature is about 6,000 K, interspersed with atomic absorption lines from the tenuous layers above the photosphere. The photosphere has a particle density of $\sim 10^{23}$ m^{d3} (this is about 0.37% of the particle number per volume of Earth's atmosphere at sea level; however, photosphere particles are electrons and protons, so the average particle in air is 58 times as heavy).

During early studies of the optical spectrum of the photosphere, some absorption lines were found that did not correspond to any chemical elements then known on Earth. In 1868, Norman Lockyer hypothesized that these absorption lines were because of a new element which he dubbed *helium*, after the Greek Sun god Helios. It was not until 25 years later that helium was isolated on Earth.

Atmosphere

The parts of the Sun above the photosphere are referred to collectively as the *solar atmosphere*. They can be viewed with telescopes operating across the electromagnetic spectrum, from radio through visible light to gamma rays, and comprise five principal zones: the *temperature minimum*, the chromosphere, the transition region, the corona, and the heliosphere. The heliosphere, which may be considered the tenuous outer atmosphere of the Sun, extends outward past the orbit of Pluto to the heliopause, where it forms a sharp shock front boundary with the interstellar medium. The chromosphere, transition region, and corona are much hotter than the surface of the Sun. The reason has not been conclusively proven; evidence suggests that Alfvén waves may have enough energy to heat the corona.

The coolest layer of the Sun is a temperature minimum region about 500 km above the photosphere, with a temperature of about 4,100 K. This part of the Sun is cool enough to support simple molecules such as carbon monoxide and water, which can be detected by their absorption spectra. Above the temperature minimum layer is a layer about 2,000 km thick, dominated by a spectrum of emission and absorption lines. It is called the *chromosphere* from the Greek root *chroma*, meaning colour, because the chromosphere is visible as a colored flash at the beginning and end of total eclipses of the Sun. The temperature in the chromosphere increases gradually with altitude, ranging up to around 20,000 K near the top. In the upper part of chromosphere helium becomes partially ionized.

Above the chromosphere, in a thin (about 200 km) transition region, the temperature rises rapidly from around 20,000 K in the upper chromosphere to coronal temperatures closer to 1,000,000 K. The temperature increase is facilitated by the full ionization of helium in the transition region, which significantly reduces radiative cooling of the plasma. The transition region does not occur at a well-defined altitude. Rather, it forms a kind of nimbus around chromospheric features such as spicules and filaments, and is in constant, chaotic motion. The transition region is not easily visible from Earth's surface, but is readily observable from space by instruments sensitive to the extreme ultraviolet portion of the spectrum.

The corona is the extended outer atmosphere of the Sun, which is much larger in volume than the Sun itself. The corona continuously expands into space forming the solar wind, which fills all the Solar System. The low corona, which is very near the surface of the Sun, has a particle density around 10^{15}–10^{16} m^{-3}. The average temperature of the corona and solar wind is about 1,000,000–2,000,000 K; however, in the hottest regions it is 8,000,000–20,000,000 K. While no complete theory yet exists to account for the temperature of the corona, at least some of its heat is known to be from magnetic reconnection.

The heliosphere, which is the cavity around the Sun filled with the solar wind plasma, extends from approximately 20 solar radii (0.1 AU) to the outer fringes of the Solar System. Its inner boundary is defined as the layer in which the flow of the solar wind becomes *superalfvénic*—that is, where the flow becomes faster than the speed of Alfvén waves. Turbulence and dynamic forces outside this boundary cannot affect the shape of the solar corona within, because the information can only travel at the speed of Alfvén waves. The solar wind travels outward continuously through the heliosphere, forming the solar magnetic field into a spiral shape, until it impacts the heliopause more than 50 AU from the Sun. In December 2004, the Voyager 1 probe passed through a shock front that is thought to be part of the heliopause. Both of the Voyager probes have recorded higher levels of energetic particles as they approach the boundary.

MAGNETIC FIELD

The Sun is a magnetically active star. It supports a strong, changing magnetic field that varies year-to-year and reverses direction about every eleven years around solar maximum. The Sun's magnetic field leads to many effects that are collectively called solar activity, including sunspots on the surface of the Sun, solar flares, and variations in solar wind that carry material through the Solar System. Effects of solar activity on Earth include auroras at moderate to high latitudes, and the disruption of radio communications and electric power. Solar activity is thought to have played a large role in the formation and evolution of the Solar System. Solar activity changes the structure of Earth's outer atmosphere.

All matter in the Sun is in the form of gas and plasma because of its high temperatures. This makes it possible for the Sun to rotate faster at its equator (about 25 days) than it does at higher latitudes (about 35 days near its poles). The differential rotation of the Sun's latitudes causes its magnetic field lines to become twisted together over time, causing magnetic field loops to erupt from the Sun's surface and trigger the formation of the Sun's dramatic sunspots and solar prominences. This twisting action creates the solar dynamo and an 11-year solar cycle of magnetic activity as the Sun's magnetic field reverses itself about every 11 years.

The solar magnetic field extends well beyond the Sun itself. The magnetized solar wind plasma carries Sun's magnetic field into the space forming what is called the interplanetary magnetic field. Since the plasma can only move along the magnetic field lines, the interplanetary magnetic field is initially stretched radially away from the Sun. Because the fields above and below the solar equator have different polarities pointing towards and away from the Sun, there exists a thin current layer in the solar equatorial plane, which is called the heliospheric current sheet. At the large distances the rotation of the Sun twists the magnetic field and the current sheet into the Archimedean spiral like structure called the Parker spiral. The interplanetary magnetic field is much stronger than the dipole component of the solar magnetic field. The Sun's 50–400 ìT (in the photosphere) magnetic dipole field reduces with the cube of the distance to about 0.1 nT at the distance of the Earth. However, according to spacecraft observations the interplanetary field at the Earth's location is about 100 times greater at around 5 nT.

CHEMICAL COMPOSITION

The Sun is composed primarily of the chemical elements hydrogen and helium; they account for 74.9% and 23.8% of the mass of the Sun in the photosphere, respectively. All heavier elements, called *metals* in astronomy, account for less than 2 percent of the mass. The most abundant metals are oxygen (roughly 1% of the Sun's mass), carbon (0.3%), neon (0.2%), and iron (0.2%).

The Sun inherited its chemical composition from the interstellar medium out of which it formed: the hydrogen and helium in the Sun were produced by Big Bang nucleosynthesis. The metals were produced by stellar nucleosynthesis in generations of stars which completed their stellar evolution and returned their material to the interstellar medium before the formation of the Sun. The chemical composition of the photosphere is normally considered representative of the composition of the primordial Solar System. However, since the Sun formed, the helium and heavy elements have settled out of the photosphere. Therefore, the photosphere now contains slightly less helium and only 84% of the heavy elements than the protostellar Sun did; the protostellar Sun was 71.1% hydrogen, 27.4% helium, and 1.5% metals.

In the inner portions of the Sun, nuclear fusion has modified the composition by converting hydrogen into helium, so the innermost portion of the Sun is now roughly 60% helium, with the metal abundance unchanged. Because the interior of the Sun is radiative, not convective, none of the fusion products from the core have risen to the photosphere.

The solar heavy-element abundances described above are typically measured both using spectroscopy of the Sun's photosphere and by measuring abundances in meteorites that have never been heated to melting temperatures. These meteorites are thought to retain the composition of the protostellar Sun and thus not affected by settling of heavy elements. The two methods generally agree well.

Singly Ionized Iron Group Elements

In the 1970s, much research focused on the abundances of iron group elements in the Sun. Although significant research was done, the abundance determination of some iron group elements (e.g., cobalt and manganese) was still difficult at least as far as 1978 because of their hyperfine structures. The first largely complete set of oscillator strengths of singly ionized iron group elements were made available first in the 1960s, and improved oscillator strengths were computed in 1976. In 1978 the abundances of singly ionized elements of the iron group were derived.

Solar and Planetary Mass Fractionation Relationship

Various authors have considered the existence of a mass fractionation relationship between the isotopic compositions of solar and planetary noble gases, for example correlations between isotopic compositions of planetary and solar neon and xenon. Nevertheless, the belief that the whole Sun has the same composition as the solar atmosphere was still widespread, at least until 1983. In 1983, it was claimed that it was the fractionation in the Sun itself that caused the fractionation relationship between the isotopic compositions of planetary and solar wind implanted noble gases.

SOLAR CYCLES

Sunspots and the Sunspot Cycle

When observing the Sun with appropriate filtration, the most immediately visible features are usually its sunspots, which are well-defined surface areas that appear darker than their surroundings because of lower temperatures. Sunspots are regions of intense magnetic activity where convection is inhibited by strong magnetic fields, reducing energy transport from the hot interior to the surface. The magnetic field causes strong heating in the corona, forming active regions that are the source of intense solar flares and coronal mass ejections. The largest sunspots can be tens of thousands of kilometres across.

The number of sunspots visible on the Sun is not constant, but varies over an 11-year cycle known as the solar cycle. At a typical solar minimum, few sunspots are visible, and occasionally none at all can be seen. Those that do appear are at high solar latitudes. As the sunspot cycle progresses, the number of sunspots increases and they move closer to the equator of the Sun, a phenomenon described by Spörer's law. Sunspots usually exist as pairs with opposite magnetic polarity. The magnetic polarity of the leading sunspot alternates every solar cycle, so that it will be a north magnetic pole in one solar cycle and a south magnetic pole in the next. The solar cycle has a great influence on space weather, and is a significant influence on the Earth's climate since luminosity has a direct relationship with magnetic activity. Solar activity minima tend to be correlated with colder temperatures, and longer than average solar cycles tend to be correlated with hotter temperatures. In the 17th century, the solar cycle appears to have stopped entirely for several decades; very few sunspots were observed during this period. During this era, which is known as the Maunder minimum or Little Ice Age, Europe experienced very cold temperatures. Earlier extended minima have been discovered through analysis of tree rings and appear to have coincided with lower-than-average global temperatures.

Possible Long-term Cycle

A recent theory claims that there are magnetic instabilities in the core of the Sun that cause fluctuations with periods of either 41,000 or 100,000 years. These could provide a better explanation of the ice ages than the Milankovitch cycles.

LIFE CYCLE

The Sun was formed about 4.57 billion years ago when a hydrogen molecular cloud collapsed. Solar formation is dated in two ways: the Sun's current main sequence age, determined using computer models of stellar evolution and nucleocosmochronology, is thought to be about 4.57 billion years. This is in close accord with the radiometric date of the oldest Solar System material, at 4.567 billion years ago.

The Sun is about halfway through its main-sequence evolution, during which nuclear fusion reactions in its core fuse hydrogen into helium. Each second, more than four million metric tons of matter are converted into energy within the Sun's core, producing neutrinos and solar radiation. At this rate, the Sun has so far converted around 100 Earth-masses of matter into energy. The Sun will spend a total of approximately 10 billion years as a main sequence star. The Sun does not have enough mass to explode as a supernova. Instead, in about 5 billion years, it will enter a red giant phase, its outer layers expanding as the hydrogen fuel in the core is consumed and the core contracts and heats up. Helium fusion will begin when the core temperature reaches around 100 million K and will produce carbon, entering the asymptotic giant branch phase.

Earth's fate is precarious. As a red giant, the Sun will have a maximum radius beyond the Earth's current orbit, 1 AU (1.5×10^{11} m), 250 times the present radius of the Sun. However, by the time it is an asymptotic giant branch star, the Sun will have lost roughly 30% of its present mass due to a stellar wind, so the orbits of the planets will move outward. If it were only for this, Earth would probably be spared, but new research suggests that Earth will be swallowed by the Sun owing to tidal interactions. Even if Earth would escape incineration in the Sun, still all its water will be boiled away and most of its atmosphere would escape into space. Even during its current life in the main sequence, the Sun is gradually becoming more luminous (about 10% every 1 billion years), and its surface temperature is slowly rising. The Sun used to be fainter in the past, which is possibly the reason life on Earth has only existed for about 1 billion years on land. The increase in solar temperatures is such that in about another billion years the surface of the Earth will likely become too hot for liquid water to exist, ending all terrestrial life.

Following the red giant phase, intense thermal pulsations will cause the Sun to throw off its outer layers, forming a planetary nebula. The only object that will remain after the outer layers are ejected is the extremely hot stellar core, which will slowly cool and fade as a white dwarf over many billions of years. This stellar evolution scenario is typical of low-to medium-mass stars.

Sunlight

Sunlight is Earth's primary source of energy. The solar constant is the amount of power that the Sun deposits per unit area that is directly exposed to sunlight. The solar constant is equal to approximately 1,368 W/m^2 (watts per square meter) at a distance of one astronomical unit (AU) from the Sun (that is, on or near Earth). Sunlight on the surface of Earth is attenuated by the Earth's atmosphere so that less power arrives at the surface—closer to 1,000 W/m^2 in clear conditions when the Sun is near the zenith.

Solar energy can be harnessed by a variety of natural and synthetic processes—photosynthesis by plants captures the energy of sunlight and converts it to chemical form (oxygen and reduced carbon compounds), while direct heating or electrical conversion by solar cells are used by solar power equipment to generate electricity or to do other useful work, sometimes employing concentrating solar power (that it is measured in suns). The energy stored in petroleum and other fossil fuels was originally converted from sunlight by photosynthesis in the distant past.

MOTION AND LOCATION WITHIN THE GALAXY

The Sun's motion about the centre of mass of the Solar System is complicated by perturbations from the planets. Every few hundred years this motion switches between prograde and retrograde. The Sun lies close to the inner rim of the Milky Way Galaxy's Orion Arm, in the Local Fluff or the Gould Belt, at a hypothesized distance of 7.5–8.5 kpc (25,000–28,000 lightyears) from the Galactic Centre, contained within the Local Bubble, a space of rarefied hot gas, possibly produced by the supernova remnant, Geminga. The distance between the local arm and the next arm out, the Perseus Arm, is about 6,500 light-years. The Sun, and thus the Solar System, is found in what scientists call the galactic habitable zone.

The Apex of the Sun's Way, or the solar apex, is the direction that the Sun travels through space in the Milky Way, relative to other nearby stars. The general direction of the Sun's galactic motion is towards the star Vega in the constellation of Lyra at an angle of roughly 60 sky degrees to the direction of the Galactic Centre.

The Sun's orbit around the Galaxy is expected to be roughly elliptical with the addition of perturbations due to the galactic spiral arms and non-uniform mass distributions. In addition the Sun oscillates up and down relative to the galactic plane approximately 2.7 times per orbit. This is very similar to how a simple harmonic oscillator works with no drag force (damping) term. It has been argued that the Sun's passage through the higher density spiral arms often coincides with mass extinctions on Earth, perhaps due to increased impact events. It takes the Solar System about 225–250 million years to complete one orbit of the galaxy (a *galactic year*), so it is thought to have completed 20–25 orbits during the lifetime of the Sun. The orbital speed of the Solar System about the centre of the Galaxy is approximately 251 km/s. At this speed, it takes around 1,190 years for the Solar System to travel a distance of 1 light-year, or 7 days to travel 1 AU.

THEORETICAL PROBLEMS

Solar Neutrino Problem

For many years the number of solar electron neutrinos detected on Earth was $^1D_3$ to $^1D_2$ of the number predicted by the standard solar model. This anomalous result was termed the solar neutrino problem. Theories proposed to resolve the problem either tried to reduce the temperature of the Sun's interior to explain the lower neutrino flux, or posited that electron neutrinos could oscillate—that is, change into undetectable tau and muon neutrinos as they traveled between the Sun and the Earth. Several neutrino observatories were built in the 1980s to measure the solar neutrino flux as accurately as possible, including the Sudbury Neutrino Observatory and Kamiokande. Results from these observatories eventually led to the discovery that neutrinos have a very small rest mass and do indeed oscillate. Moreover, in 2001 the Sudbury Neutrino Observatory was able to detect all three types of neutrinos directly, and found that the Sun's *total* neutrino emission rate agreed with the Standard Solar Model, although depending on the neutrino energy as few as one-third of the neutrinos seen at Earth are of the electron type. This proportion agrees with that predicted by the Mikheyev-Smirnov-Wolfenstein effect (also known as the matter effect), which describes neutrino oscillation in matter, and it is now considered a solved problem.

Coronal Heating Problem

The optical surface of the Sun (the photosphere) is known to have a temperature of approximately 6,000 K. Above it lies the solar corona, rising to a temperature of 1,000,000–2,000,000 K. The high temperature of the corona shows that it is heated by something other than direct heat conduction from the photosphere. It is thought that the energy necessary to heat the corona is provided by turbulent motion in the convection zone below the photosphere, and two main mechanisms have been proposed to explain coronal heating. The first is wave heating, in which sound, gravitational or magnetohydrodynamic waves are produced by turbulence in the convection zone. These waves travel upward and dissipate in the corona, depositing their energy in the ambient gas in the form of heat. The other is magnetic heating, in which magnetic energy is continuously built up by photospheric motion and released through magnetic reconnection in the form of large solar flares and myriad similar but smaller events—nanoflares.

Currently, it is unclear whether waves are an efficient heating mechanism. All waves except Alfvén waves have been found to dissipate or refract before reaching the corona. In addition, Alfvén waves do not easily dissipate in the corona. Current research focus has therefore shifted towards flare heating mechanisms.

Faint young Sun Problem

Theoretical models of the Sun's development suggest that 3.8 to 2.5 billion years ago, during the Archean period, the Sun was only about 75% as bright as it is today. Such a weak star would not have been able to sustain liquid water on the Earth's surface, and thus life should not have been able to develop.

However, the geological record demonstrates that the Earth has remained at a fairly constant temperature throughout its history, and that the young Earth was somewhat warmer than it is today. The consensus among scientists is that the young Earth's atmosphere contained much larger quantities of greenhouse gases (such as carbon dioxide, methane and/or ammonia) than are present today, which trapped enough heat to compensate for the smaller amount of solar energy reaching the planet.

Present Anomalies

The Sun is currently behaving unexpectedly in a number of ways.

- It is in the midst of an unusual sunspot minimum, lasting far longer and with a higher percentage of spotless days than normal; since May 2008.
- It is measurably dimming; its output has dropped 0.02% at visible wavelengths and 6% at EUV wavelengths in comparison with the levels at the last solar minimum.
- Over the last two decades, the solar wind's speed has dropped by 3%, its temperature by 13%, and its density by 20%.
- Its magnetic field is at less than half strength compared to the minimum of 22 years ago. The entire heliosphere, which fills the Solar System, has shrunk as a result, resulting in an increase in the level of cosmic radiation striking the Earth and its atmosphere.

HISTORY OF OBSERVATION

Early Understanding and Etymology

The English proper noun *sun* developed from Old English *sunne* (around 725, attested in *Beowulf*), and may be related to *south*. Cognates to English *sun* appear in other Germanic languages, including Old Frisian *sunne*, *sonne* ("sun"), Old Saxon *sunna*, Middle Dutch *sonne*, modern Dutch *zon*, Old High German *sunna*, modern German *Sonne*, Old Norse *sunna*, and Gothic *sunnô*. All Germanic terms for the Sun stem from Proto-Germanic **sunnôn*. In Germanic paganism, the Sun is personified as a goddess; Sól/Sunna.

Theories have been proposed that Sun, as Germanic goddess, may represent an extension of an earlier Proto-Indo-European deity due to Indo-European linguistic connections between Norse *Sól*, Sanskrit *Surya*, Gaulish *Sulis*, Lithuanian *Saulë*, and Slavic *Solnitse*.

Humanity's most fundamental understanding of the Sun is as the luminous disk in the sky, whose presence above the horizon creates day and whose absence causes night. In many prehistoric and ancient cultures, the Sun was thought to be a solar deity or other supernatural phenomenon. Worship of the Sun was central to civilizations such as the Inca of South America and the Aztecs of what is now Mexico. Many ancient monuments were constructed with solar phenomena in mind; for example, stone megaliths accurately mark the summer or winter solstice (some of the most prominent megaliths are located in Nabta Playa, Egypt, Mnajdra, Malta and at Stonehenge, England); Newgrange, a prehistoric human-built mount in Ireland, was designed to detect the winter solstice; the pyramid of El Castillo at Chichén Itzá in Mexico is designed to cast shadows in the shape of serpents climbing the pyramid at the vernal and autumn equinoxes.

In the late Roman Empire the Sun's birthday was a holiday celebrated as Sol Invictus (literally "unconquered sun") soon after the winter solstice which may have been an antecedent to Christmas. Regarding the fixed stars, the Sun appears from Earth to revolve once a year along the ecliptic through the zodiac, and so Greek astronomers considered it to be one of the seven planets (Greek *planetes*, "wanderer"), after which the seven days of the week are named in some languages.

Development of Scientific Understanding

In the early first millennium B.C.E., Babylonian astronomers observed that the Sun's motion along the ecliptic was not uniform, though they were unaware of why this was; it is today known that this is due to the Earth moving in an elliptic orbit around the Sun, with the Earth moving faster when it is nearer to the Sun at perihelion and moving slower when it is farther away at aphelion.

One of the first people to offer a scientific or philosophical explanation for the Sun was the Greek philosopher Anaxagoras, who reasoned that it was a giant flaming ball of metal even larger than the Peloponnesus rather than the chariot of Helios, and that the Moon reflected the light of the Sun. For teaching this heresy, he was imprisoned by the authorities and sentenced to death, though he was later released through the intervention of Pericles. Eratosthenes estimated the distance between the Earth and the Sun in the 3rd century BCE as "of stadia myriads 400 and 80000", the translation of which is ambiguous, implying either 4,080,000 stadia (755,000 km) or 804,000,000 stadia (148 to 153 million kilometres); the latter value is correct to within a few percent. In the 1st century CE, Ptolemy estimated the distance as 1,210 times the Earth radius.

The theory that the Sun is the centre around which the planets move was first proposed by the ancient Greek Aristarchus of Samos in the 3rd century BCE, and later adopted by Seleucus of Seleucia. This largely philosophical view was developed into fully predictive mathematical model of a heliocentric system in the 16th century by Nicolaus Copernicus. In the early 17th century, the invention of the telescope permitted detailed observations of sunspots by Thomas Harriot, Galileo Galilei and other astronomers. Galileo made some of the first known telescopic observations of sunspots and posited that they were on the surface of the Sun rather than small objects passing between the Earth and the Sun. Sunspots were also observed since the Han Dynasty (206 BCE – 220 CE) by Chinese astronomers who maintained records of these observations for centuries. Averroes also provided a description of sunspots in the 12th century.

Arabic astronomical contributions include Albatenius discovering that the direction of the Sun's eccentric is changing, and Ibn Yunus observing more than 10,000 entries for the Sun's position for many years using a large astrolabe. The transit of Venus was first observed in 1032 by Avicenna, who concluded that Venus is closer to the Earth than the Sun, while one of the first observations of the transit of Mercury was conducted by Ibn Bajjah in the 12th century.

In 1672 Giovanni Cassini and Jean Richer determined the distance to Mars and were thereby able to calculate the distance to the Sun. Isaac Newton observed the Sun's light using a prism, and showed that it was made up of light of many colors, while in 1800 William Herschel discovered infrared radiation beyond the red part of the solar spectrum. The 1800s saw spectroscopic studies of the Sun advance, and Joseph von Fraunhofer made the first observations of absorption lines in the spectrum, the strongest of which are still often referred to as Fraunhofer lines. When expanding the spectrum of light from the Sun, a large number of missing colors can be found.

In the early years of the modern scientific era, the source of the Sun's energy was a significant puzzle. Lord Kelvin suggested that the Sun was a gradually cooling liquid body that was radiating an internal store of heat. Kelvin and Hermann von Helmholtz then proposed a gravitational contraction mechanism to explain the energy output. Unfortunately the resulting age estimate was only 20 million years, well short of the time span of at least 300 million years suggested by some geological discoveries of that time. In 1890 Joseph Lockyer, who discovered helium in the solar spectrum, proposed a meteoritic hypothesis for the formation and evolution of the Sun.

Not until 1904 was a documented solution offered. Ernest Rutherford suggested that the Sun's output could be maintained by an internal source of heat, and suggested radioactive decay as the source. However, it would be Albert Einstein who would provide the essential clue to the source of the Sun's energy output with his mass-energy equivalence relation $E = mc^2$.

In 1920, Sir Arthur Eddington proposed that the pressures and temperatures at the core of the Sun could produce a nuclear fusion reaction that merged hydrogen (protons) into helium nuclei, resulting in a production of energy from the net change in mass. The preponderance of hydrogen in the Sun was confirmed in 1925 by Cecilia Payne. The theoretical concept of fusion was developed in the 1930s by the astrophysicists Subrahmanyan Chandrasekhar and Hans Bethe. Hans Bethe calculated the details of the two main energy-producing nuclear reactions that power the Sun.

Finally, a seminal paper was published in 1957 by Margaret Burbidge, entitled "Synthesis of the Elements in Stars". The paper demonstrated convincingly that most of the elements in the universe had been synthesized by nuclear reactions inside stars, some like our Sun.

Solar Space Missions

The first satellites designed to observe the Sun were NASA's Pioneers 5, 6, 7, 8 and 9, which were launched between 1959 and 1968. These probes orbited the Sun at a distance similar to that of the Earth, and made the first detailed measurements of the solar wind and the solar magnetic field. Pioneer 9 operated for a particularly long time, transmitting data until May 1983.

In the 1970s, two Helios spacecraft and the Skylab Apollo Telescope Mount provided scientists with significant new data on solar wind and the solar corona. The Helios 1 and 2 probes were U.S.–German collaborations that studied the solar wind from an orbit carrying the spacecraft inside Mercury's orbit at perihelion. The Skylab space station, launched by NASA in 1973, included a solar observatory module called the Apollo Telescope Mount that was operated by astronauts resident on the station. Skylab made the first time-resolved observations of the solar transition region and of ultraviolet emissions from the solar corona. Discoveries included the first observations of coronal mass ejections, then called "coronal transients", and of coronal holes, now known to be intimately associated with the solar wind.

In 1980, the Solar Maximum Mission was launched by NASA. This spacecraft was designed to observe gamma rays, X-rays and UV radiation from solar flares during a time of high solar activity and solar luminosity. Just a few months after launch, however, an electronics failure caused the probe to go into standby mode, and it spent the next three years in this inactive state. In 1984 Space Shuttle Challenger mission STS-41C retrieved the satellite and repaired its electronics before re-releasing it into orbit. The Solar Maximum Mission subsequently acquired thousands of images of the solar corona before re-entering the Earth's atmosphere in June 1989.

Launched in 1991, Japan's Yohkoh (*Sunbeam*) satellite observed solar flares at X-ray wavelengths. Mission data allowed scientists to identify several different types of flares, and demonstrated that the corona away from regions of peak activity was much more dynamic and active than had previously been supposed. Yohkoh observed an entire solar cycle but went into standby mode when an annular eclipse in 2001 caused it to lose its lock on the Sun. It was destroyed by atmospheric re-entry in 2005.

One of the most important solar missions to date has been the Solar and Heliospheric Observatory, jointly built by the European Space Agency and NASA and launched on 2 December 1995. Originally intended to serve a two-year mission, a mission extension through 2012 was approved in October 2009.

It has proven so useful that a follow-on mission, the Solar Dynamics Observatory, was launched in February 2010. Situated at the Lagrangian point between the Earth and the Sun (at which the gravitational pull from both is equal), SOHO has provided a constant view of the Sun at many wavelengths since its launch. Besides its direct solar observation, SOHO has enabled the discovery of a large number of comets, mostly very tiny sungrazing comets which incinerate as they pass the Sun.

All these satellites have observed the Sun from the plane of the ecliptic, and so have only observed its equatorial regions in detail. The Ulysses probe was launched in 1990 to study the Sun's polar regions.

It first travelled to Jupiter, to "slingshot" past the planet into an orbit which would take it far above the plane of the ecliptic. Serendipitously, it was well-placed to observe the collision of Comet Shoemaker-Levy 9 with Jupiter in 1994. Once Ulysses was in its scheduled orbit, it began observing the solar wind and magnetic field strength at high solar latitudes, finding that the solar wind from high latitudes was moving at about 750 km/s which was slower than expected, and that there were large magnetic waves emerging from high latitudes which scattered galactic cosmic rays.

Elemental abundances in the photosphere are well known from spectroscopic studies, but the composition of the interior of the Sun is more poorly understood. A solar wind sample return mission, Genesis, was designed to allow astronomers to directly measure the composition of solar material. Genesis returned to Earth in 2004 but was damaged by a crash landing after its parachute failed to deploy on re-entry into Earth's atmosphere. Despite severe damage, some usable samples have been recovered from the spacecraft's sample return module and are undergoing analysis.

The Solar Terrestrial Relations Observatory (STEREO) mission was launched in October 2006. Two identical spacecraft were launched into orbits that cause them to (respectively) pull further ahead of and fall gradually behind the Earth. This enables stereoscopic imaging of the Sun and solar phenomena, such as coronal mass ejections.

OBSERVATION AND EFFECTS

Sunlight is very bright, and looking directly at the Sun with the naked eye for brief periods can be painful, but is not particularly hazardous for normal, non-dilated eyes. Looking directly at the Sun causes phosphene visual artifacts and temporary partial blindness. It also delivers about 4 milliwatts of sunlight to the retina, slightly heating it and potentially causing damage in eyes that cannot respond properly to the brightness. UV exposure gradually yellows the lens of the eye over a period of years and is thought to contribute to the formation of cataracts, but this depends on general exposure to solar UV, not on whether one looks directly at the Sun. Long-duration viewing of the direct Sun with the naked eye can begin to cause UV-induced, sunburn-like lesions on the retina after about 100 seconds, particularly under conditions where the UV light from the Sun is intense and well focused; conditions are worsened by young eyes or new lens implants (which admit more UV than aging natural eyes), Sun angles near the zenith, and observing locations at high altitude.

Viewing the Sun through light-concentrating optics such as binoculars is very hazardous without an appropriate filter that blocks UV and substantially dims the sunlight. An attenuating (ND) filter might not filter UV and so is still dangerous. Attenuating filters to view the Sun should be specifically designed for that use: some improvised filters pass UV or IR rays that can harm the eye at high brightness levels.

Unfiltered binoculars can deliver over 500 times as much energy to the retina as using the naked eye, killing retinal cells almost instantly (even though the power per unit area of image on the retina is the same, the heat cannot dissipate fast enough because the image is larger). Even brief glances at the midday Sun through unfiltered binoculars can cause permanent blindness.

Partial solar eclipses are hazardous to view because the eye's pupil is not adapted to the unusually high visual contrast: the pupil dilates according to the total amount of light in the field of view, *not* by the brightest object in the field.

During partial eclipses most sunlight is blocked by the Moon passing in front of the Sun, but the uncovered parts of the photosphere have the same surface brightness as during a normal day. In the overall gloom, the pupil expands from ~2 mm to ~6 mm, and each retinal cell exposed to the solar image receives about ten times more light than it would looking at the non-eclipsed Sun.

This can damage or kill those cells, resulting in small permanent blind spots for the viewer. The hazard is insidious for inexperienced observers and for children, because there is no perception of pain: it is not immediately obvious that one's vision is being destroyed.

During sunrise and sunset sunlight is attenuated due to Rayleigh scattering and Mie scattering from a particularly long passage through Earth's atmosphere, and the Sun is sometimes faint enough to be viewed comfortably with the naked eye or safely with optics (provided there is no risk of bright sunlight suddenly appearing through a break between clouds). Hazy conditions, atmospheric dust, and high humidity contribute to this atmospheric attenuation.

A rare optical phenomenon may occur shortly after sunset or before sunrise, known as a green flash. The flash is caused by light from the Sun just below the horizon being bent (usually through a temperature inversion) towards the observer. Light of shorter wavelengths (violet, blue, green) is bent more than that of longer wavelengths (yellow, orange, red) but the violet and blue light is scattered more, leaving light that is perceived as green. Ultraviolet light from the Sun has antiseptic properties and can be used to sanitize tools and water. It also causes sunburn, and has other medical effects such as the production of vitamin D. Ultraviolet light is strongly attenuated by Earth's ozone layer, so that the amount of UV varies greatly with latitude and has been partially responsible for many biological adaptations, including variations in human skin colour in different regions of the globe.

Terminology

Like other natural phenomena, the Sun has been an object of veneration in many cultures throughout human history, and was the source of the word Sunday. Its formal name in the English language is, per the International Astronomical Union, the Sun (capitalized as a proper noun). The Latin name *Sol*, for the Sun god of the same name in English, is widely known but not common in general English language use; the adjectival form is the related word *solar*. "Sol" is, however, the modern word for "Sun" in many European languages. The term *sol* is also used by planetary astronomers to refer to the duration of a solar day on another planet, such as Mars. A mean Earth solar day is approximately 24 hours, while a mean Martian 'sol' is 24 hours, 39 minutes, and 35.244 seconds.

SUNLIGHT

Sunlight, in the broad sense, is the total frequency spectrum of electromagnetic radiation given off by the Sun. On Earth, sunlight is filtered through the Earth's atmosphere, and solar radiation is obvious as daylight when the Sun is above the horizon. When the direct solar radiation is not blocked by clouds, it is experienced as sunshine, a combination of bright light and radiant heat. When it is blocked by the clouds or reflects off of other objects, it is experienced as diffused light.

The World Meteorological Organization uses the term "sunshine duration" to mean the cumulative time during which an area receives direct irradiance from the Sun of at least 120 watts per square meter. Sunlight may be recorded using a sunshine recorder, pyranometer or pyrheliometer. Sunlight takes about 8.3 minutes to reach the Earth.

Direct sunlight has a luminous efficacy of about 93 lumens per watt of radiant flux, which includes infrared, visible, and ultraviolet light. Bright sunlight provides illuminance of approximately 100,000 lux or lumens per square meter at the Earth's surface. Sunlight is a key factor in photosynthesis, a process vital for life on Earth.

Calculation

To calculate the amount of sunlight reaching the ground, both the elliptical orbit of the Earth and the attenuation by the

Earth's atmosphere have to be taken into account. The extraterrestrial solar illuminance (E_{ext}), corrected for the elliptical orbit by using the day number of the year (dn), is given by

$$E_{sc} = \frac{E_{ext}}{1 + 0.033412 \cdot \cos\left(2\pi \frac{dn-3}{365}\right)};$$

where dn=1 on January 1; dn=2 on January 2; dn=32 on February 1, etc. In this formula dn-3 is used, because in modern times Earth's perihelion, the closest approach to the Sun and therefore the maximum E_{ext} occurs around January 3 each year. The value of.033412 is determined knowing that the ratio between the perihelion (0.98328989 AU) squared and the aphelion (1.01671033 AU) squared should be approximately 0.935338.

The solar illuminance constant (E_{sc}), is equal to 128×10^3 lx. The direct normal illuminance (E_{dn}), corrected for the attenuating effects of the atmosphere is given by:

$$E_{dn} = E_{ext}\, e^{-cm};$$

where c is the atmospheric extinction coefficient and m is the relative optical airmass.

Solar Constant

The solar constant, a measure of flux, is the amount of incoming solar electromagnetic radiation per unit area that would be incident on a plane perpendicular to the rays, at a distance of one astronomical unit (AU) (roughly the mean distance from the Sun to the Earth). When solar irradiance is measured on the outer surface of Earth's atmosphere, the measurements can be adjusted using the inverse square law to infer the magnitude of solar irradiance at one AU and deduce the solar constant. The solar constant includes all types of solar radiation, not just the visible light. It is measured by satellite to be roughly 1.366 kilowatts per square meter (kW/m^2).

Sunlight Intensity in the Solar System

Different bodies of the Solar System receive light of an intensity inversely proportional to the square of their distance from Sun. A rough table comparing the amount of light received by each planet on the Solar System follows:

Planet	*Perihelion-Aphelion distance (AU)*	*Solar radiation maximum and minimum (W/m²)*
Mercury	0.3075 – 0.4667	14,446 – 6,272
Venus	0.7184 – 0.7282	2,647 – 2,576
Earth	0.9833 – 1.017	1,413 – 1,321
Mars	1.382 – 1.666	715 – 492
Jupiter	4.950 – 5.458	55.8 – 45.9
Saturn	9.048 – 10.12	16.7 – 13.4
Uranus	18.38 – 20.08	4.04 – 3.39
Neptune	29.77 – 30.44	1.54 – 1.47

The actual brightness of sunlight that would be observed at the surface depends also on the presence and composition of an atmosphere.

For example Venus' thick atmosphere reflects more than 60% of the solar light it receives. The actual illumination of the surface is about 14,000 lux, comparable to that on Earth "in the daytime with overcast clouds". Sunlight on Mars would be more or less like daylight on Earth wearing sunglasses, and as can be seen in the pictures taken by the rovers, there is enough diffuse sky radiation that shadows would not seem particularly dark. Thus it would give perceptions and "feel" very much like Earth daylight.

For comparison purposes, sunlight on Saturn is slightly brighter than Earth sunlight at the average sunset or sunrise. Even on Pluto the sunlight would still be bright enough to almost match the average living room. To see sunlight as dim as full moonlight on the Earth, a distance of about 500 AU (~69 light-hours) is needed; there is only a handful of objects in the solar system known to orbit farther than such a distance, among them 90377 Sedna and (87269) 2000 OO67.

Composition

The spectrum of the Sun's solar radiation is close to that of a black body with a temperature of about 5,800 K. About half that lies in the visible short-wave part of the electromagnetic spectrum and the other half mostly in the near-infrared part. Some also lies in the ultraviolet part of the spectrum. When ultraviolet radiation is not absorbed by the atmosphere or other protective coating, it can cause damage to the skin known as sunburn or trigger an adaptive change in human skin pigmentation. The spectrum of electromagnetic radiation striking the Earth's atmosphere is 100 to 10^6 nanometers (nm). This can be divided into five regions in increasing order of wavelengths:

- Ultraviolet C or (UVC) range, which spans a range of 100 to 280 nm. The term *ultraviolet* refers to the fact that the radiation is at higher frequency than violet light (and, hence also invisible to the human eye). Owing to absorption by the atmosphere very little reaches the Earth's surface (Lithosphere). This spectrum of radiation has germicidal properties, and is used in germicidal lamps.
- Ultraviolet B or (UVB) range spans 280 to 315 nm. It is also greatly absorbed by the atmosphere, and along with UVC is responsible for the photochemical reaction leading to the production of the Ozone layer.
- Ultraviolet A or (UVA) spans 315 to 400 nm. It has been traditionally held as less damaging to the DNA, and hence used in tanning and PUVA therapy for psoriasis.
- Visible range or light spans 400 to 700 nm. As the name suggests, it is this range that is visible to the naked human eye.
- Infrared range that spans 700 nm to 10^6 nm [1 (mm)]. It is responsible for an important part of the electromagnetic radiation that reaches the Earth. It is also divided into three types on the basis of wavelength:
  * Infrared-A: 700 nm to 1,400 nm
  * Infrared-B: 1,400 nm to 3,000 nm
  * Infrared-C: 3,000 nm to 1 mm.

Surface Illumination

The spectrum of surface illumination depends upon solar elevation due to atmospheric effects, with the blue spectral component from atmospheric scatter dominating during twilight before and after sunrise and sunset, respectively, and red dominating during sunrise and sunset. These effects are apparent in natural light photography where the principal source of illumination is sunlight as mediated by the atmosphere.

According to Craig Bohren, "preferential absorption of sunlight by ozone over long horizon paths gives the zenith sky its blueness when the sun is near the horizon".

Climate Effects

On Earth, solar radiation is obvious as daylight when the sun is above the horizon. This is during daytime, and also in summer near the poles at night, but not at all in winter near the poles. When the direct radiation is not blocked by clouds, it is experienced as *sunshine*, combining the perception of bright white light (sunlight in the strict sense) and warming.

The warming on the body, the ground and other objects depends on the absorption (electromagnetic radiation) of the electromagnetic radiation in the form of heat.

The amount of radiation intercepted by a planetary body varies inversely with the square of the distance between the star and the planet. The Earth's orbit and obliquity change with time (over thousands of years), sometimes forming a nearly perfect circle, and at other times stretching out to an orbital eccentricity of 5% (currently 1.67%). The total insolation remains almost constant due to Kepler's second law,

$$\tfrac{2A}{r^2}\,dt = d\theta,$$

where A is the "areal velocity" invariant. That is, the integration over the orbital period (also invariant) is a constant.

$$\int_0^T \tfrac{2A}{r^2}\,dt = \int_0^{2\pi} d\theta = \text{constant}$$

If we assume the solar radiation power P as a constant over time and the solar irradiation given by the inverse-square law, we obtain also the average insolation as a constant.

But the seasonal and latitudinal distribution and intensity of solar radiation received at the Earth's surface also varies. For example, at latitudes of 65 degrees the change in solar energy in summer & winter can vary by more than 25% as a result of the Earth's orbital variation.

Because changes in winter and summer tend to offset, the change in the annual average insolation at any given location is near zero, but the redistribution of energy between summer and winter does strongly affect the intensity of seasonal cycles. Such changes associated with the redistribution of solar energy are considered a likely cause for the coming and going of recent ice ages.

Life on Earth

The existence of nearly all life on Earth is fueled by light from the sun. Most autotrophs, such as plants, use the energy of sunlight, combined with carbon dioxide and water, to produce simple sugars—a process known as photosynthesis.

These sugars are then used as building blocks and in other synthetic pathways which allow the organism to grow. Heterotrophs, such as animals, use light from the sun indirectly by consuming the products of autotrophs, either directly or by consuming other heterotrophs. The sugars and other molecular components produced by the autotrophs are then broken down, releasing stored solar energy, and giving the heterotroph the energy required for survival. This process is known as respiration.

In prehistory, humans began to further extend this process by putting plant and animal materials to other uses. They used animal skins for warmth, for example, or wooden weapons to hunt. These skills allowed humans to harvest more of the sunlight than was possible through glycolysis alone, and human population began to grow.

During the Neolithic Revolution, the domestication of plants and animals further increased human access to solar energy. Fields devoted to crops were enriched by inedible plant matter, providing sugars and nutrients for future harvests. Animals which had previously only provided humans with meat and tools once they were killed were now used for labour throughout their lives, fueled by grasses inedible to humans.

The more recent discoveries of coal, petroleum and natural gas are modern extensions of this trend. These fossil fuels are the remnants of ancient plant and animal matter, formed using energy from sunlight and then trapped within the earth for millions of years.

Because the stored energy in these fossil fuels has accumulated over many millions of years, they have allowed modern humans to massively increase the production and consumption of primary energy. As the amount of fossil fuel is large but finite, this cannot continue indefinitely, and various theories exist as to what will follow this stage of human civilization (e.g. alternative fuels, Malthusian catastrophe, new urbanism, peak oil).

Cultural Aspects

Many people find direct sunlight to be too bright for comfort, especially when reading from white paper upon which the sun is directly shining. Indeed, looking directly at the sun can cause long-term vision damage.

To compensate for the brightness of sunlight, many people wear sunglasses. Cars, many helmets and caps are equipped with visors to block the sun from direct vision when the sun is at a low angle. In colder countries, many people prefer sunnier days and often avoid the shade. In hotter countries the converse is true; during the midday hours many people prefer to stay inside to remain cool. If they do go outside, they seek shade which may be provided by trees, parasols, and so on. Sunshine is often blocked from entering buildings through the use of walls, window blinds, awnings, shutters or curtains.

Sunbathing

Sunbathing is a popular leisure activity in which a person sits or lies in direct sunshine. People often sunbathe in comfortable places where there is ample sunlight. Some common places for sunbathing include beaches, open air swimming pools, parks, gardens, and sidewalk cafés. Sunbathers typically wear limited amounts of clothing or some simply go nude. An alternative some use to sunbathing is to use a sunbed that generates ultraviolet light and can be used indoors regardless of outdoor weather conditions and amount of sun light.

For many people with pale or brownish skin, one purpose for sunbathing is to darken one's skin colour (get a sun tan) as this is considered in some cultures to be beautiful, associated with outdoor activity, vacations/holidays, and health. Some people prefer nude sunbathing so that an "all-over" or "even" tan can be obtained.

Skin tanning is achieved by an increase in the dark pigment inside skin cells called melanocytes and it is actually an automatic response mechanism of the body to sufficient exposure to ultraviolet radiation from the sun or from artificial sunlamps. Thus, the tan gradually disappears with time, when one is no longer exposed to these sources.

EFFECTS ON HUMAN HEALTH

The body produces vitamin D from sunlight (specifically from the UVB band of ultraviolet light), and excessive seclusion from the sun can lead to deficiency unless adequate amounts are obtained through diet. This very common deficiency leaves the body generally vulnerable to cancers.

On the other hand, excessive sunlight exposure has been linked to all types of skin cancer caused by the ultraviolet part of radiation from sunlight or sunlamps.

Sunburn can have mild to severe inflammation effects on skin; this can be avoided by using a proper sunscreen cream or lotion or by gradually building up melanocytes with increasing exposure. Another detrimental effect of UV exposure is accelerated skin aging (also called skin photodamage), which produces a difficult to treat cosmetic effect. Some people are concerned that ozone depletion is increasing the incidence of such health hazards. A 10% decrease in ozone could cause a 25% increase in skin cancer.

A lack of sunlight, on the other hand, is considered one of the primary causes of seasonal affective disorder (SAD), a serious form of the "winter blues". SAD occurrence is more prevalent in locations further from the tropics, and most of the treatments (other than prescription drugs) involve light therapy, replicating sunlight via lamps tuned to specific (visible, not ultra-violet) wavelengths of light or full-spectrum bulbs.

A recent study indicates that more exposure to sunshine early in a person's life relates to less risk from multiple sclerosis (MS) later in life.

SOLAR OBSERVATORY

A solar observatory is an observatory that specializes in monitoring the Sun. As such, they usually have one or more solar telescopes. Solar observatories study phenomena associated with the Sun. The Sun, being the closest star to earth, allows a unique chance to study stellar physics with high-resolution. It was, until the 1990's, the only star whose surface had been resolved. General topics that interest a solar astronomer are its 11-year periodicity (i.e., the Solar Cycle), sunspots, magnetic field activity, solar flares, coronal mass ejections, differential rotation, and plasma physics.

12

Future of the Earth

The future of the Earth will be determined by a variety of factors, including increases in the luminosity of the Sun, loss of heat energy from the Earth's core, perturbations by the other bodies in the Solar System and the biochemistry at the Earth's surface. Milankovitch theory predicts the planet will continue to undergo glaciation cycles because of eccentricity, axial tilt, and precession of the Earth's orbit. As part of the ongoing supercontinent cycle, plate tectonics will probably result in a supercontinent in 250 million–350 million years. Some time in the next 1.5 billion–4.5 billion years, the axial tilt of the Earth may begin to undergo chaotic variations, with changes in the axial tilt of up to 90°. One billion to two billion years in the future, the steady increase in solar radiation caused by the helium build-up at the core of the Sun will result in the loss of the oceans and the cessation of continental drift. Four billion years from now, the increase in the Earth's surface temperature will cause a runaway greenhouse effect. By that point, most if not all the life on the surface will be extinct. The most likely ultimate fate of the planet is absorption by the Sun in about 7.5 billion years, after the star has entered the red giant phase and expanded to cross the planet's orbit.

HUMAN INFLUENCE

Humans now play a key role in the biosphere, with the large human population dominating many of Earth's ecosystems. This has resulted in a widespread, ongoing extinction of other species during the present geological epoch, now known as the Holocene extinction. The large scale loss of species caused by human influence since the 1950s has been called a biotic crisis, with an estimated 10% of the total species lost as of 2007. At current rates, about 30% of species are at risk of extinction in the next hundred years. The Holocene extinction event is the result of habitat destruction, the widespread distribution of invasive species, hunting, and climate change. In the present day, human activity has had a significant impact on the surface of the planet. More than a third of the land surface has been modified by human actions, and humans use about 20% of global primary production. The concentration of carbon dioxide in the atmosphere has increased by close to 30% since the start of the Industrial Revolution.

The consequences of a persistent biotic crisis have been predicted to last for at least five million years. It could result in a decline in biodiversity and homogenization of biotas, accompanied by a proliferation of species that are opportunistic, such as pests and weeds. Novel species may also emerge; in particular taxa that prosper in human-dominated ecosystems may rapidly diversify into many new species. Microbes are likely to benefit from the increase in nutrient-enriched environmental niches. However, no new species of existing large vertebrates are likely to arise and food chains will probably be shorter.

ORBIT AND ROTATION

The gravitational perturbations of the other planets in the Solar System combine to modify the orbit of the Earth and the orientation of its spin axis. These changes can influence the planetary climate.

Glaciation

Historically, there have been cyclical periods of glaciation in which ice sheets covered the higher latitudes of the continents. The Milankovitch theory predicts that glaciation occurs because of astronomical factors in combination with climate feedback mechanisms and plate tectonics. The primary astronomical drivers are a higher than normal orbital eccentricity, a low axial tilt (or obliquity), and the alignment of summer solstice with the aphelion. Each of these effects occur cyclically. For example, the eccentricity changes over time cycles of about

100,000 and 400,000 years, with the value ranging from less than 0.01 up to 0.05. This is equivalent to a change of the semiminor axis of the planet's orbit from 99.95% of the semimajor axis to 99.88%, respectively.

At present the Earth is in an interglacial period, which would normally be expected to end in about 25,000 years. The current rate of increased carbon dioxide release into the atmosphere by humans may delay the onset of the next period of glaciation until at least 50,000–130,000 years from now. However, a global warming period of finite duration (based on the assumption that fossil fuel use will cease by the year 2200) will probably only impact the glaciation cycle for about 5,000 years. Thus, a brief period of global warming induced through a few centuries worth of greenhouse gas emission would only have a limited impact in the long term.

Obliquity

The tidal acceleration of the Moon slows the rotation rate of the Earth and increases the Earth-Moon distance. Other effects that can dissipate the Earth's rotational energy are friction between the core and mantle, tides in the atmosphere, convection in the mantle, and climate changes that can increase or decrease the ice load at the poles. These combined effects are expected to increase the length of the day by more than 1.5 hours over the next 250 million years, and to increase the obliquity by about a half degree. The distance to the Moon will increase by about 1.5 Earth radii during the same period.

Based on computer models, the presence of the Moon appears to stabilize the obliquity of the Earth, which may help the planet to avoid dramatic climate changes. This stability is achieved because the Moon increases the precession rate of the Earth's spin axis, thereby avoiding resonances between the precession of the spin and precession frequencies of the ascending node of the planet's orbit. (That is, the precession motion of the ecliptic.) However, as the semimajor axis of the Moon's orbit continues to increase in the future, this stabilizing effect will diminish. At some point perturbation effects will probably cause chaotic variations in the obliquity of the Earth, and the axial tilt may change by angles as high as 90° from the plane of the orbit. This is expected to occur within about

1.5–4.5 billion years, although the exact time is unknown. A high obliquity would probably result in dramatic changes in the climate and may destroy the planet's habitability. When the axial tilt of the Earth reaches 54°, the equator will receive less radiation from the Sun than the poles. The planet could remain at an obiliquity of 60° to 90° for periods as long as 10 million years.

Plate Tectonics

The theory of plate tectonics demonstrates that the continents of the Earth are moving across the surface at the rate of a few centimeters per year. This is expected to continue, causing the plates to relocate and collide. Continental drift is facilitated by two factors: the energy generation within the planet and the presence of a hydrosphere. With the loss of either of these, continental drift will come to a halt. The production of heat through radiogenic processes is sufficient to maintain mantle convection and plate subduction for at least the next 1.1 billion years.

At present, the continents of North and South America are moving westward from Africa and Europe. Researchers have produced several scenarios about how this will continue in the future. These geodynamic models can be distinguished by the subduction flux, whereby the oceanic crust moves under a continent. In the introversion model, the younger, interior, Atlantic ocean becomes preferentially subducted and the current migration of North and South America is reversed. In the extroversion model, the older, exterior, Pacific ocean remains preferentially subducted and North and South America migrate toward eastern Asia.

As the understanding of geodynamics improves, these models will be subject to revision. In 2008, for example, a computer simulation was used to predict that a reorganization of the mantle convection will occur, causing a supercontinent to form around Antarctica.

Regardless of the outcome of the continental migration, the continued subduction process causes water to be transported to the mantle. After a billion years from the present, a geophysical model gives an estimate that 27% of the current ocean mass will have been subducted. If this process were to continue unmodified into the future, the subduction and release would reach a point of stability after 65% of the current ocean mass has been subducted.

Introversion

Christopher Scotese and his colleagues have mapped out the predicted motions several hundred million years into the future as part of the Paleomap Project. In their scenario, 50 million years from now the Mediterranean sea may vanish and the collision between Europe and Africa will create a long mountain range extending to the current location of the Persian Gulf. Australia will merge with Indonesia, and Baja California will slide northward along the coast. New subduction zones may appear off the eastern coast of North and South America, and mountain chains will form along those coastlines. To the south, the migration of Antarctica to the north will cause all of its ice sheets to melt. This, along with the melting of the Greenland ice sheets, will raise the average ocean level by 90 metres (300 ft). The inland flooding of the continents will result in climate changes.

As this scenario continues, by 100 million years from the present the continental spreading will have reached its maximum extent and the continents will then begin to coalesce. In 250 million years, North America will collide with Africa while South America will wrap around the southern tip of Africa. The result will be the formation of a new supercontinent (sometimes called Pangaea Ultima), with the Pacific Ocean stretching across half the planet. The continent of Antarctica will reverse direction and return to the South Pole, building up a new ice cap.

Extroversion

The first scientist to extrapolate the current motions of the continents was Canadian geologist Paul F. Hoffman of Harvard University. In 1992, Hoffman predicted that continents of North and South America would continue to advance across the Pacific Ocean, pivoting about Siberia until they begin to merge with Asia. He dubbed the resulting supercontinent, Amasia. Later, in the 1990s, Roy Livermore calculated a similar scenario. He predicted that Antarctica would start to migrate northward, and east Africa and Madagascar would move across the Indian Ocean to collide with Asia.

In an extroversion model, the closure of the Pacific Ocean would be complete by about 350 million years. This marks the completion of the current supercontinent cycle, wherein the continents split apart and then rejoin each other about every 400–500 million years. Once the supercontinent is built, plate tectonics may enter a period of inactivity as the rate of subduction drops by an order of magnitude. This period of stability could cause an increase in the mantle temperature at the rate of 30–100 K every 100 million years, which is the minimum lifetime of past supercontinents. As a consequence, volcanic activity may increase.

Supercontinent

The formation of a supercontinent can dramatically affect the environment. The collision of plates will result in mountain building, thereby shifting weather patterns. Sea levels may fall because of increased glaciation. The rate of surface weathering can rise, resulting in an increase in the rate that organic material is buried. Supercontinents can cause a drop in global temperatures and an increase in atmospheric oxygen. These changes can result in more rapid biological evolution as new niches emerge. This, in turn, can affect the climate, further lowering temperatures.

The formation of a supercontinent insulates the mantle. The flow of heat will be concentrated, resulting in volcanism and the flooding of large areas with basalt. Rifts will form and the supercontinent will split up once more.

SOLAR EVOLUTION

The energy generation of the Sun is based upon thermonuclear fusion of hydrogen into helium. This occurs in the core region of the star using the proton–proton chain reaction process. Because there is no convection in the solar core, the fusion process results in a steady build-up of helium. The temperature at the core of the Sun is too low for nuclear fusion of helium atoms through the triple-alpha process, so these atoms do not contribute to the net energy generation that is needed to maintain hydrostatic equilibrium of the Sun.

At present, nearly half the hydrogen at the core has been consumed, with the remainder consisting primarily of helium. To compensate for the steadily decreasing number of hydrogen atoms per unit mass, the core temperature of the Sun has gradually increased through a rise in pressure. This has caused the remaining hydrogen to undergo fusion at a more rapid rate, thereby generating the energy needed to maintain the equilibrium. The result has been a steady increase in the energy output of the Sun. This increase can be approximated by the formula:

$$L(t) = \left[1 + \frac{2}{5}\left(1 - \frac{t}{t_{Sun}}\right)\right]^{-1} L_{Sun}$$

where t is a time period less than or equal to the present time t_{Sun}, $L(t)$ is the luminosity at time t, and L_{Sun} is the current solar luminosity.

When the Sun first became a main sequence star, it radiated only 70% of the current luminosity. The luminosity has increased in a nearly linear fashion to the present, increasing by 1% every 110 million years. Likewise, in three billion years the Sun is expected to be 33% more luminous. The hydrogen fuel at the core will finally be exhausted in 4.8 billion years, when the Sun will be 67% more luminous than at present. Thereafter the Sun will continue to burn hydrogen in a shell surrounding its core, until the increase in luminosity reaches 121% of the present value. This marks the end of the Sun's main sequence lifetime, and thereafter it will evolve into a red giant.

Climate Impact

As the global temperature of the Earth climbs because of the rising luminosity of the Sun, the rate of weathering of silicate minerals will increase. This in turn will decrease the level of carbon dioxide in the atmosphere. Within the next 600 million years from the present, the concentration of CO_2 will fall below the critical threshold needed to sustain C_3 photosynthesis: about 50 parts per million. At this point, trees and forests in their current forms will no longer be able to survive. However, C_4 carbon fixation can continue at much lower concentrations, down to above 10 parts per million. Thus plants using C_4 photosynthesis may be able to survive for at least 0.8 billion years and possibly as long as 1.2 billion years from now, after which rising temperatures will make the biosphere unsustainable. Currently, C_4 plants represent about 5% of Earth's plant biomass and 1% of its known plant species. For example, about 50% of all grass species (Poaceae) use the C_4 photosynthetic pathway, as do many species in the herbaceous family Amaranthaceae.

When the levels of carbon dioxide fall to the limit where photosynthesis is barely sustainable, the proportion of carbon dioxide in the atmosphere is expected to oscillate up and down. This will allow land vegetation to reappear each time the level of carbon dioxide rises due to tectonic activity and animal life. However, the long term trend is for the plant life on land to die off altogether as most of the remaining carbon in the atmosphere becomes sequestered in the Earth. Some microbes are capable of photosynthesis at concentrations of CO_2 of a few parts per million, so these life forms would probably disappear only because of rising temperatures and the loss of the biosphere.

In their work *The Life and Death of Planet Earth*, authors Peter D. Ward and Donald Brownlee have argued that some form of animal life may continue even after most of the Earth's plant life has disappeared. Initially, they expect that some insects, lizards, birds and small mammals may persist, along with sea life. Without oxygen replenishment by plant life, however, they believe that the animals would probably die off from asphyxiation within a few million years. Even if sufficient oxygen were to remain in the atmosphere through the persistence of some form of photosynthesis, the steady rise in global temperature would result in a gradual loss of biodiversity. Much of the surface would become a barren desert and life would primarily be found in the oceans.

Once the solar luminosity is 10% higher than its current value, the average global surface temperature reaches 320 K (47 °C). The atmosphere will become a humid greenhouse leading to a runaway evaporation of the oceans. At this point, models of the Earth's future environment demonstrate that the stratosphere would contain increasing levels of water. These water molecules will be broken down through photodissociation by solar ultraviolet radiation, allowing hydrogen to escape the atmosphere. The net result would be a loss of the world's sea water in about 1.1 billion years from the present.

Still, there will continue to be some reservoirs at the surface as water is steadily released from the deep crust and mantle. Some water may be retained at the poles and there may be occasional rainstorms, but for the most part the planet would be a dry desert. What happens next depends on the level of tectonic activity. The release of carbon dioxide by volcanic eruption may eventually cause the atmosphere to enter a "supergreenhouse" state like that of the planet Venus. However, without surface water, plate tectonics would probably come to a halt and most of the carbonates would remain securely buried.

The loss of the oceans could be delayed until two billion years in the future if the total atmospheric pressure were to decline. A lower atmospheric pressure would reduce the greenhouse effect, thereby lowering the surface temperature. This could occur if natural processes were to remove the nitrogen from the atmosphere. Studies of organic sediments has shown that at least 100 kilopascals (1 bar) of nitrogen has been removed from the atmosphere over the past four billion years; enough to effectively double the current atmospheric pressure if it were to be released. This rate of removal would be sufficient to counter the effects of increasing solar luminosity for the next two billion years. However, beyond that point, the amount of water in the lower atmosphere will have risen to 40% and the runaway moist greenhouse will commence.

A runaway greenhouse effect will take place when the luminosity from the Sun reaches 40% more than its current value, four billion years from now. The atmosphere will heat up and the surface temperature will rise. However, most of the atmosphere will be retained until the Sun has entered the red giant stage.

Red Giant Stage

Once the Sun changes from burning hydrogen at the core to burning hydrogen around a shell, the core will start to contract and the outer envelope will expand. The total luminosity will steadily increase over the next billion years until it reaches 2,730 times the Sun's current luminosity at the age of 12.167 billion years. During this phase the Sun will undergo mass loss, with about 33% of its total mass shed with the solar wind. The loss of mass will mean that the orbits of the planets will expand. The orbital distance of the Earth will increase to at most 150% of its current value.

The most rapid part of the Sun's expansion into a red giant occurs during the final stages, when the Sun is about 12 billion years old. It is likely to expand to swallow both Mercury and Venus, reaching a maximum radius of 1.2 astronomical units (180 Gm). The Earth will interact tidally with the Sun's outer atmosphere, which would serve to decrease the orbital radius. Drag from the chromosphere of the Sun would also reduce the Earth's orbit. These effects will act to counterbalance the mass loss by the Sun, and the Earth will most likely be engulfed by the sun.

By the time the Sun begins to grow as a red giant, the orbit of the Moon will have expanded until it takes 47 days to complete. The drag from the solar atmosphere may cause the orbit of the Moon to decay. Once the orbit of the Moon closes to a distance of 18,470 km, it will cross the Earth's Roche limit. The tidal interaction with the Earth will break apart the Moon, turning it into a ring system. Most of the orbiting ring will then begin to decay, and the debris will impact the Earth. Hence, even if the Earth is not swallowed up by the Sun, the planet may be left moonless.

13

Figure of the Earth

The expression figure of the Earth has various meanings in geodesy according to the way it is used and the precision with which the Earth's size and shape is to be defined. The actual topographic surface is most apparent with its variety of land forms and water areas. This is, in fact, the surface on which actual Earth measurements are made. It is not suitable, however, for exact mathematical computations, because the formulas which would be required to take the irregularities into account would necessitate a prohibitive amount of computations. The topographic surface is generally the concern of topographers and hydrographers.

The Pythagorean concept of a spherical Earth offers a simple surface which is mathematically easy to deal with. Many astronomical and navigational computations use it as a surface representing the Earth. While the sphere is a close approximation of the true figure of the Earth and satisfactory for many purposes, to the geodesists interested in the measurement of long distances—spanning continents and oceans—a more exact figure is necessary. Closer approximations range from modelling the shape of the entire Earth as an oblate spheroid or an oblate ellipsoid, to the use of spherical harmonics or local approximations in terms of local reference ellipsoids. The idea of a planar or flat surface for Earth, however, is still acceptable for surveys of small areas, as local topography is more important than the curvature. Plane-table surveys are made for relatively small areas, and no account is taken of the curvature of the Earth. A survey of a city would likely be computed as though the Earth were a plane surface the size of the city. For such small areas, exact positions can be determined relative to each other without considering the size and shape of the total Earth.

In the mid-to late-20th century, research across the geosciences contributed to drastic improvements in the accuracy of the Figure of the Earth. The primary utility (and the motivation for funding, mainly from the military) of this improved accuracy was to provide geographical and gravitational data for the inertial guidance systems of ballistic missiles. This funding also drove the expansion of geoscientific disciplines, fostering the creation and growth of various geoscience departments at many universities.

ELLIPSOID OF REVOLUTION

Since the Earth is flattened at the poles and bulging at the equator, the geometrical figure used in geodesy to most nearly approximate Earth's shape is an oblate spheroid. An oblate spheroid, or oblate ellipsoid, is an ellipsoid of revolution obtained by rotating an ellipse about its shorter axis. A spheroid describing the figure of the Earth or other celestial body is called a reference ellipsoid. The reference ellipsoid for the Earth is called Earth ellipsoid.

An ellipsoid of revolution is uniquely defined by two numbers— two dimensions, or one dimension and a number representing the difference between the two dimensions. Geodesists, by convention, use the semimajor axis and flattening. The size is represented by the radius at the equator—the semimajor axis of the cross-sectional ellipse—and designated by the letter a. The shape of the ellipsoid is given by the flattening, f, which indicates how much the ellipsoid departs from spherical. (In practice, the two defining numbers are usually the equatorial radius and the reciprocal of the flattening, rather than the flattening itself; for the WGS84 spheroid used by today's GPS systems, the reciprocal of the flattening is set at 298.257223563 exactly.)

The difference between a sphere and a reference ellipsoid for Earth is small, only about one part in 300. Historically flattening was computed from grade measurements. Nowadays geodetic networks and satellite geodesy are used. In practice, many reference ellipsoids have been developed over the centuries from different surveys. The flattening value varies slightly from one reference ellipsoid to another, reflecting local conditions and whether the reference ellipsoid is intended to model the entire earth or only some portion of it.

A sphere has a single radius of curvature, which is simply the radius of the sphere. More complex surfaces have radii of curvature that vary over the surface. The radius of curvature describes the radius of the sphere that best approximates the surface at that point. Oblate ellipsoids have constant radius of curvature east to west along parallels, if a graticule is drawn on the surface, but varying curvature in any other direction. For an oblate ellipsoid, the polar radius of curvature r_p is larger than the equatorial

$$r_p = \frac{a^2}{b}:$$

because the pole is flattened: the flatter the surface, the larger the sphere must be to approximate it. Conversely, the ellipsoid's north-south radius of curvature at the equator r_e is smaller than the polar

$$r_e = \frac{b^2}{a}.$$

HISTORICAL EARTH ELLIPSOIDS

The reference ellipsoid models listed below have had utility in geodetic work and many are still in use. The older ellipsoids are named for the individual who derived them and the year of development is given. In 1887 the English mathematician Col Alexander Ross Clarke CB FRS RE was awarded the Gold Medal of the Royal Society for his work in determining the figure of the Earth. The international ellipsoid was developed by John Fillmore Hayford in 1910 and adopted by the International Union of Geodesy and Geophysics (IUGG) in 1924, which recommended it for international use.

At the 1967 meeting of the IUGG held in Lucerne, Switzerland, the ellipsoid called GRS-67 (Geodetic Reference System 1967) in the listing was recommended for adoption.

The new ellipsoid was not recommended to replace the International Ellipsoid (1924), but was advocated for use where a greater degree of accuracy is required. It became a part of the GRS-67 which was approved and adopted at the 1971 meeting of the IUGG held in Moscow. It is used in Australia for the Australian Geodetic Datum and in South America for the South American Datum 1969.

The GRS-80 (Geodetic Reference System 1980) as approved and adopted by the IUGG at its Canberra, Australia meeting of 1979 is based on the equatorial radius (semi-major axis of Earth ellipsoid) a, total mass GM, dynamic form factor J_2 and angular velocity of rotation ù, making the inverse flattening $1/f$ a derived quantity. The minute difference in $1/f$ seen between GRS-80 and WGS-84 results from an unintentional truncation in the latter's defining constants: while the WGS-84 was designed to adhere closely to the GRS-80, incidentally the WGS-84 derived flattening turned out to be slightly different than the GRS-80 flattening because the normalized second degree zonal harmonic gravitational coefficient, that was derived from the GRS-80 value for J2, was truncated to 8 significant digits in the normalization process.

An ellipsoidal model describes only the ellipsoid's geometry and a normal gravity field formula to go with it. Commonly an ellipsoidal model is part of a more encompassing geodetic datum. For example, the older ED-50 (European Datum 1950) is based on the Hayford or International Ellipsoid. WGS-84 is peculiar in that the same name is used for both the complete geodetic reference system and its component ellipsoidal model. Nevertheless the two concepts—ellipsoidal model and geodetic reference system—remain distinct.

MORE COMPLICATED FIGURES

The possibility that the Earth's equator is an ellipse rather than a circle and therefore that the ellipsoid is triaxial has been a matter of scientific controversy for many years. Modern technological developments have furnished new and rapid methods for data collection and since the launch of *Sputnik 1*, orbital data have been used to investigate the theory of ellipticity.

A second theory, more complicated than triaxiality, proposed that observed long periodic orbital variations of the first Earth satellites indicate an additional depression at the south pole accompanied by a bulge of the same degree at the north pole. It is also contended that the northern middle latitudes were slightly flattened and the southern middle latitudes bulged in a similar amount. This concept suggested a slightly pear-shaped Earth and was the subject of much public discussion. Modern geodesy tends to retain the ellipsoid of revolution and treat triaxiality and pear shape as a part of the geoid figure: they are represented by the spherical harmonic coefficients C_{22},S_{22} and C_{30}, respectively, corresponding to degree and order numbers 2.2 for the triaxiality and 3.0 for the pear shape.

Geoid

It was stated earlier that measurements are made on the apparent or topographic surface of the Earth and it has just been explained that computations are performed on an ellipsoid. One other surface is involved in geodetic measurement: the geoid. In geodetic surveying, the computation of the geodetic coordinates of points is commonly performed on a reference ellipsoid closely approximating the size and shape of the Earth in the area of the survey. The actual measurements made on the surface of the Earth with certain instruments are however referred to the geoid. The ellipsoid is a mathematically defined regular surface with specific dimensions. The geoid, on the other hand, coincides with that surface to which the oceans would conform over the entire Earth if free to adjust to the combined effect of the Earth's mass attraction (gravitation) and the centrifugal force of the Earth's rotation. As a result of the uneven distribution of the Earth's mass, the geoidal surface is irregular and, since the ellipsoid is a regular surface, the separations between the two, referred to as geoid undulations, geoid heights, or geoid separations, will be irregular as well.

The geoid is a surface along which the gravity potential is everywhere equal and to which the direction of gravity is always perpendicular. The latter is particularly important because optical instruments containing gravity-reference leveling devices are commonly used to make geodetic measurements. When properly adjusted, the vertical axis of the instrument coincides with the direction of gravity and is, therefore, perpendicular to the geoid. The angle between the plumb line which is perpendicular to the geoid (sometimes called "the vertical") and the perpendicular to the ellipsoid (sometimes called "the ellipsoidal normal") is defined as the deflection of the vertical. It has two components: an east-west and a north-south component.

Earth Rotation and Earth's Interior

Determining the exact figure of the Earth is not only a geodetic operation or a task of geometry, but is also related to geophysics. Without any idea of the Earth's interior, we can state a "constant density" of 5.515 g/cm³ and, according to theoretical arguments, such a body rotating like the Earth would have a flattening of 1:230.

In fact the measured flattening is 1:298.25, which is more similar to a sphere and a strong argument that the Earth's core is *very compact*. Therefore the density must be a function of the depth, reaching from about 2.7 g/cm³ at the surface up to approximately 15 within the inner core. Modern seismology yields a value of 16 g/cm³ at the centre of the earth.

Global and Regional Gravity Field

Also with implications for the physical exploration of the Earth's interior is the gravitational field, which can be measured very accurately at the surface and remotely by satellites. True vertical generally does not correspond to theoretical vertical (deflection ranges from 2– to 50–) because topography and all *geological masses* disturb the gravitational field. Therefore the gross structure of the earth's crust and mantle can be determined by geodetic-geophysical models of the subsurface.

Bibliography

Adams, Simon; David Lambert: *Earth Science: An Illustrated Guide to Science*, New York, Chelsea House, 2006.

Allaby M.: *Dictionary of Earth Sciences*, Oxford University Press, 2008.

Andrews, Edgar H.: *God, Science and Evolution,* Presbyterian and Reformed, 1980.

Austin, Steven A.: *Catastrophes in Earth History,* Institute for Creation Research, 1984.

Barnes, Thomas G.: *Origin and Destiny of the Earth's Magnetic Field,* Creation Research Society, New York, 1983.

Bhanu, T.K. Chandra : *Earth Science and Technology*, Campus Books, Delhi, 2009.

Bliss, Richard B.: *Voyage to the Stars,* Institute for Creation Research, New York, 1991.

Chamberlin, T. C.: *The Methods of the Earth Science*, Popular Science Monthly, 1904.

Cloud, P.: *Oasis in Space: Earth History from the Beginning*, New York, Norton Books, 1988.

Crang, Mike; Thrift, Nigel J.: *Thinking Space*, London: Routledge, 2000.

Ghosh, Parthasarathi : *Numerical Methods and Models in Earth Science*, New India Publishing Agency, Delhi, 2011.

Hallam, A.: *A Revolution in the Earth Sciences*, New York, Oxford University Press, 1973.

Howe, George F.: *Speak to the Earth,* Creation Research Society, New York, 1975.

Krebs, Robert E : *The Basics of Earth Science*, Atlantic, Delhi, 2007.

Kumar, Sunil : *Environmental Change Management*, ACB Pub, Delhi, 2004.

Kusky, Timothy : *Viva-Facts on File: The Encyclopedia of Earth Science*, Viva Books, Delhi, 2006.

Lahiri, Dipankar : *Samsad Encyclopaedic Dictionary of Earth Sciences*, Shishu Sahitya Samsad, Delhi, 2004.

Oldroyd, David: *Earth Cycles: A Historical Perspective*, Westport, Connicticut: Greenwood Press, 2006.

Prasad, Govind : *Environmental Geomorphology*, Discovery Publishing House, Delhi, 2008.

Prasad, K.N. : *An Introduction to Earth Science: Heritage Sites*, APH, Delhi, 2006.

Reed, Christina: *Earth Science: Decade by Decade*, New York, Facts on File, 2008.

Sahu, B.K. : *Statistical Models in Earth Sciences*, B S Pub, Delhi, 2005.

Sarma, D.D. : *Geostatistics with Applications in Earth Sciences*, Capital Pub, Delhi, 2002.

Sharma, V K : *Origin and Development of Earth Science*, Vista International Pub, Delhi, 2008.

Simon, Adams; David Lambert: *Earth Science: An Illustrated Guide to Science*, New York, 2006.

Singh, Chandra Prakash : *Applied Geomorphology : A Study*, B R Pub, Delhi, 2002.

Smith, Gary A.; Aurora Pun: *How Does the Earth Work?*. Upper Saddle River, NJ, 2006.

Stiegeler, Stella : *Universities Press Dictionary of Earth Science*, Universities Press, Delhi, 2003.

Tripathi, Satish C. : *Advances in Earth Sciences*, Satish Serial Publishing House, Delhi, 2009.

Index